跟戴铭学
iOS编程
理顺核心知识点

戴铭 著

电子工业出版社
Publishing House of Electronics Industry
北京·BEIJING

内 容 简 介

本书针对 iOS 应用开发涉及的核心知识点进行详细剖析，分为 Swift、编译器、iOS 中的大前端技术三章，从应用、进阶、未来三方面剖析 iOS 开发。第 1 章从 Swift 源码和第三方库源码层面来分析 iOS 开发在实际工作中的常用核心知识，包括数据类型、编程范式、内存管理、网络处理、页面布局、完整的转译器和解释器等内容。第 2 章通过分析 LLVM 源码、编译后的可执行文件内部结构、链接器来讲解 iOS 开发进阶知识，同时介绍在实际工作中的应用场景和示例。第 3 章介绍在未来的 iOS"大前端浪潮"中需要掌握的核心知识，包括 JavaScriptCore、WebCore 等。

本书适合对 iOS 有兴趣的开发人员学习，也适合经验丰富的 iOS 开发者和对编程语言本身有兴趣的人员参考。

未经许可，不得以任何方式复制或抄袭本书之部分或全部内容。
版权所有，侵权必究。

图书在版编目（CIP）数据

跟戴铭学 iOS 编程：理顺核心知识点 / 戴铭著. —北京：电子工业出版社，2020.1
ISBN 978-7-121-35688-9

Ⅰ. ①跟… Ⅱ. ①戴… Ⅲ. ①移动终端－应用程序－程序设计 Ⅳ. ①TN929.53

中国版本图书馆 CIP 数据核字(2018)第 280900 号

责任编辑：官　杨
印　　刷：三河市龙林印务有限公司
装　　订：三河市龙林印务有限公司
出版发行：电子工业出版社
　　　　　北京市海淀区万寿路 173 信箱　邮编 100036
开　　本：720×1000　1/16　印张：18.75　字数：410 千字
版　　次：2020 年 1 月第 1 版
印　　次：2020 年 5 月第 3 次印刷
定　　价：79.00 元

凡所购买电子工业出版社图书有缺损问题，请向购买书店调换。若书店售缺，请与本社发行部联系，联系及邮购电话：(010) 88254888，88258888。
质量投诉请发邮件至 zlts@phei.com.cn，盗版侵权举报请发邮件至 dbqq@phei.com.cn。
本书咨询联系方式：010-51260888-819，faq@phei.com.cn。

专家推荐

孙源　滴滴出行技术专家：

底层原理和源码分析是工程师进阶的必修课，但枯燥的学习过程往往让人望而却步。本书以生动的形式，抽丝剥茧般地拨开编译器、解释器的迷雾，让读者从会使用一门语言开始，做到真正理解语言的本质并破开"黑盒"。相信认真读完此书的读者们定会有醍醐灌顶的感觉。

唐巧　资深 iOS 开发者、《iOS 开发进阶》作者：

这是一本对 iOS 开发的核心知识进行综述的图书。整本书的内容由浅入深，从泛型等 Swift 核心概念讲起，最终深入到了编译器和大前端技术等进阶内容。初级的开发者可以从中快速回顾 Swift 核心知识，需要提高的开发者则可以从编译器开始，在 iOS 开发中找到可以深入研究的各个方向。另外，本书作者亲自绘制的插画也为图书内容增添了不少趣味。总之，这是一本指引开发者们一步步提升开发技能的图书，建议大家阅读。

王巍（喵神）　LINE 高级软件工程师：

本书通过 Swift 语言作为切入点，介绍了包括用 Swift 实现语言转译器及解释器、clang 及 LLVM 分析、跨平台工作原理等高端技术内容。本书适合那些想要深入学习开发技术的开发者们阅读。

目录

第 1 章 Swift 1
   1.1 Swift 简介 1
   1.2 泛型 2
      1.2.1 类型约束 4
      1.2.2 关联类型 4
      1.2.3 类型擦除 5
      1.2.4 Where 语句 5
      1.2.5 泛型和 Any 类型 6
   1.3 集合 6
      1.3.1 基本概念 6
      1.3.2 迭代器 7
      1.3.3 Sequence 协议 8
      1.3.4 Collection 协议 9
      1.3.5 Map 9
      1.3.6 flatMap 10
      1.3.7 Reduce 12
      1.3.8 Array 13
      1.3.9 弱引用的 Swift 数组 14
      1.3.10 Dictionary 14
   1.4 协议式编程 15
   1.5 链式编程 18
   1.6 Swift 内存管理 20
      1.6.1 内存分配 20
      1.6.2 Swift 派发机制 21
      1.6.3 基本数据类型内存管理 23
      1.6.4 struct 内存管理 23

         1.6.5 class 内存管理 23
         1.6.6 协议类型内存管理 24
         1.6.7 泛型的内存管理 24
   1.7 JSON 数据的处理 24
         1.7.1 使用 JSONDecoder 24
         1.7.2 CodingKey 协议 26
         1.7.3 JSONDecoder 的 keyDecodingStrategy 属性 27
         1.7.4 枚举定义 block 31
         1.7.5 inout 31
   1.8 网络请求 32
   1.9 自动布局 SnapKit 库分析 36
         1.9.1 给谁做约束 37
         1.9.2 如何设置约束 38
         1.9.3 设置完约束后如何处理 45
   1.10 用 Swift 实现一个简单的语言转译器 46
         1.10.1 转译器简介 46
         1.10.2 词法分析器 47
         1.10.3 语法分析器 51
         1.10.4 遍历器 54
         1.10.5 转换器 54
         1.10.6 代码生成器 55
         1.10.7 Scheme 的其他特性 57
         1.10.8 Babel 57
   1.11 用 Swift 开发一个简单的解释器 61
         1.11.1 四则运算 61
         1.11.2 算术表达式 71
         1.11.3 中间表示 79
         1.11.4 变量 85
         1.11.5 属性 92
         1.11.6 静态检查 95

第 2 章 编译器 103
   2.1 LLVM 简介 103
   2.2 编译流程 103

- 2.3 使用 clang 命令编译 .m 文件 106
  - 2.3.1 构建 Target 107
  - 2.3.2 Target 在构建过程中的控制 107
- 2.4 clang static analyzer 108
- 2.5 IR 代码 115
  - 2.5.1 IR 结构 115
  - 2.5.2 LLVM IR 优化 118
  - 2.5.3 SSA 119
- 2.6 clang 前端组件 122
  - 2.6.1 库的介绍 122
  - 2.6.2 使用 libclang 进行语法分析 122
- 2.7 Driver 124
  - 2.7.1 Driver 的工作流程 125
  - 2.7.2 Parse 127
  - 2.7.3 Pipeline 130
  - 2.7.4 Action 130
  - 2.7.5 Bind 132
  - 2.7.6 Translate 132
  - 2.7.7 Jobs 132
  - 2.7.8 Execute 133
- 2.8 clang attribute 134
- 2.9 clang 警告处理 137
- 2.10 通过 LibTooling 控制语法树 137
- 2.11 clang 插件 138
- 2.12 LLVM Backend 138
  - 2.12.1 CodeGen 阶段 139
  - 2.12.2 SelectionDAG 140
  - 2.12.3 Register Allocation 144
  - 2.12.4 Code Emission 146
- 2.13 LLVM 优化 147
- 2.14 Swift 编译 148
- 2.15 编译后生成的二进制内容 Link Map File 149
- 2.16 编译后生成的 dSYM 文件 149
- 2.17 Mach-O 文件 150

- 2.18 如何利用 Mach-O 150
  - 2.18.1 打印堆栈信息，保存现场 151
  - 2.18.2 通过 hook 获取更多信息的方法 154
  - 2.18.3 hook msgsend 方法 159
  - 2.18.4 统计方法调用频次 161
  - 2.18.5 找出 CPU 使用的线程堆栈 163
  - 2.18.6 Demo 165
- 2.19 dyld 165
- 2.20 LLVM 工具链 168
  - 2.20.1 获取 LLVM 168
  - 2.20.2 编译 LLVM 的源代码 169
  - 2.20.3 LLVM 源代码工程目录介绍 171
  - 2.20.4 lib 目录介绍 171
  - 2.20.5 工具链命令介绍 171

第 3 章 iOS 中的大前端技术 173

- 3.1 大前端技术简介 173
- 3.2 Weex 实现技术 173
  - 3.2.1 将 iOS 工程集成 WeexSDK 173
  - 3.2.2 自定义端内能力的 Module 175
  - 3.2.3 读取用 JavaScript 写的 Weex 内容 178
  - 3.2.4 从 Vue 代码到 JS bundle 179
  - 3.2.5 在端内运行 JS bundle 的原理 179
- 3.3 JavaScriptCore 181
  - 3.3.1 JavaScriptCore 介绍 181
  - 3.3.2 JavaScriptCore 全貌 181
  - 3.3.3 JavaScriptCore 与 WebCore 184
  - 3.3.4 词法、语法分析 184
  - 3.3.5 从代码到 JIT 的过程 185
  - 3.3.6 分层编译 186
  - 3.3.7 类型分析 190
  - 3.3.8 指令集架构 192
  - 3.3.9 JavaScript 196
- 3.4 WebCore 197

- 3.4.1 浏览器历史 197
- 3.4.2 WebKit 全貌 197
- 3.4.3 WTF 203
- 3.4.4 Loader 219
- 3.4.5 HTML 词法解析 244
- 3.4.6 HTML 语法解析 247
- 3.4.7 构建 DOM Tree 249
- 3.4.8 CSS 256
- 3.4.9 RenderObject Tree 265
- 3.4.10 Layout 273

第 1 章
Swift

1.1 Swift 简介

使用 Swift 这门编程语言可以开发 iOS、macOS、Linux server-side、云服务，甚至 TensorFlow 的程序。这门语言包含了很多现在开发者喜欢的语言特性，通过引入这些先进的语法概念，Swift 可以使代码编写更加简洁，更容易编写出思路新颖的代码。这些特性包含了闭包和函数指针的统一、多返回值、泛型、支持方法、扩展和协议的结构体、函数编程模式等。苹果公司希望 Swift 能够成为 C 语言、C++ 和 Objective-C 的继承者，所以既让它具有了像类型、控制流和运算符一样的 Low-Level Primitives 特性，又让它具备了面向对象的一些高级特性。例如 Module 提供了命名空间，去掉了头文件，并且提供了 do、guard、defer 和 repeat 等具有创意的语法关键字和规则。

Swift 是一种多范式（Multi-paradigm）的编程语言，可以使用面向对象和函数式的方式来写程序，如面向协议编程、直接操作内存等。像 C 语言那样进行低层级的位操作，但是这种操作在编译环节不容易检查、不安全、容易出错，所以这种操作又具有"unsafe"特性，一般只在与 C 语言交互时使用。

Swift 作为一个强类型语言，在编译时推断完类型之后，变量和参数等都会有确定的类型，并且在编译时有着比 Objective-C 更严格的静态类型检查机制。这样的设计是为了使代码更加安全。为了安全，Swift 要求变量在使用前必须初始化。Swift 的编译器会阻止生成并避免使用 nil 对象。如果使用 nil 对象是必要的，那么可以使用 Swift 中 optional 功能。optional 里包含了 nil 对象。Swift 定义了一个"?"语法，该语法会在编译时强制开发者处理值为 nil 的情况，让你确认已安全处理了这个值。通过 LLVM 编译器优化，Swift 的构建速度非常快。

Swift 已经开源，源代码、bug 追踪器和邮件列表都能在 Swift.org 上找到。这个开源项目包含了一系列的子项目，包括 Swift 编译器命令行工具、标准库、高级功能的核心库、Swift REPL 的 LLDB 调试器，以及用于分发和构建源代码的包管理工具（Swift Package Manager）。开源后的 Swift 获得了更多的贡献者，使得它能在更多平台和技术社区里运用起来，并且能够移植到更多的非苹果公司的平台上。

1.2 泛型

泛型的概念最早出自 C++ 的模板，Swift 的泛型和 C++ 模板设计的思路是一致的。为什么不与 Java 的泛型一致呢？原因是 C++ 是一种编译时多态技术，而 Java 是运行时多态技术。Java 运行时多态是在运行时才能确定的，所以会有运行时额外的计算，缺点是通过类型擦除生成的代码只有一份。而 C++ 在编译时通过编译器来确定类型，所以在运行时就不需要额外计算了，这样效率会高一些，缺点是生成的机器码的二进制包会大一些，虽然执行快但可能会有更多的 I/O。Swift 采用编译时多态技术一方面是和 C++ 一样在 I/O 性能和执行效率中选择了执行效率，另一方面是为了让代码更加安全。当 Java 在类型擦除中进行向下转型时丢失的类型只有在运行时才能看到。而 Swift 提供了额外的类型约束，使得泛型在定义的时候就能够判定类型。

使用泛型编写代码可以让我们写出更简洁、安全的代码。类型转换的代码也可以由编译器完成，减少了手写类型强制转换带来的运行时出错的问题。随着编译器或虚拟机技术的不断进步，在代码不变的情况下，优化工作会在编译层面完成，从而获得潜在性能方面的收益。综上所述，泛型可以写出可重用、支持任意类型的函数，以及灵活、清晰、优雅的代码。

下面我们用示例分析泛型是如何解决具体问题的。

```
let dragons = ["red dragon", "white dragon", "blue dragon"]
func showDragons(dragons : [String]) {
    for dragon in dragons {
        print("\(dragon)")
    }
}
showDragons(dragons: dragons)
```

如果用数字类型作为编号来代表每只龙，并且需要一个函数来显示出它们的数字编号，那么需要再写一个函数，如下所示。

```
let dragons = [1276, 8737, 1173]
func showDragons(dragons : [Int]) {
```

```
    for dragon in dragons {
        print("\(dragon)")
    }
}
showDragons(dragons: dragons)
```

以上两个函数里的内容基本一样,只是参数的类型不同,此时就是使用泛型的最佳时刻。在 Swift 中,泛型能够在 function、class、struct、enums、protocol 和 extension 中使用。首先看一下 function 是怎么使用泛型的,如下所示。

```
let dragonsId = [1276, 8737, 1173]
let dragonsName = ["red dragon", "blue dragon", "black dragon"]
func showDragons<T>(dragons : [T]) {
    for dragon in dragons {
        print("\(dragon)")
    }
}
showDragons(dragons: dragonsName)
showDragons(dragons: dragonsId)
```

上面的类型参数 T 最好能够具有一定的描述性,就像字典(Dictionary)定义中的 key 和 value,以及数组(Array)里的 element 一样具有描述性。Swift 的基础库大量使用泛型,例如数组和字典都是泛型集合。我们既可以创建 Swift 支持的任意类型的数组,也可以创建存储任意类型值的字典。下面我们看一看系统提供的 swap 方法是如何运用泛型的,代码在 Swift 的源代码路径 stdlib/public/core/MutableCollection.swift 里,如下所示。

```
// the legacy swap free function
//
/// Exchanges the values of the two arguments.
///
/// The two arguments must not alias each other. To swap two elements of
/// a mutable collection, use the 'swapAt(_:_:)' method of that collection
/// instead of this function.
///
/// - Parameters:
///   - a: The first value to swap.
///   - b: The second value to swap.
@inlinable
public func swap<T>(_ a: inout T, _ b: inout T) {
    // Semantically equivalent to (a,b) = (b,a).
    // Microoptimized to avoid retain/release traffic.
    let p1 = Builtin.addressof(&a)
    let p2 = Builtin.addressof(&b)
    _debugPrecondition(
      p1 != p2,
      "swapping a location with itself is not supported")

    // Take from P1.
    let tmp: T = Builtin.take(p1)
    // Transfer P2 into P1.
    Builtin.initialize(Builtin.take(p2) as T, p1)
    // Initialize P2.
```

```
    Builtin.initialize(tmp, p2)
}
```

这里的两个参数 a 和 b 使用了同样的泛型 T，这样能够保障两个参数在交换后对于后续的处理在类型上是安全的。

1.2.1 类型约束

HTN 项目里状态机的 Transition 结构体的定义，如下所示。

```
struct HTNTransition<S: Hashable, E: Hashable> {
    let event: E
    let fromState: S
    let toState: S

    init(event: E, fromState: S, toState: S) {
        self.event = event
        self.fromState = fromState
        self.toState = toState
        if fromState == toState {
            print("Two state is same")
        }
    }
}
```

以上示例中的 event、fromState 和 toState 可以是不同类型的数据，也可以是枚举、字符串或者整数等。定义 S 和 E 两个不同的泛型可以让状态和事件的类型不同，这样接口会更加灵活，更容易适配更多的项目。

大家可以注意到 S 和 E 都遵循 Hashable 协议，这就是要求它们要符合这个协议的类型约束。使用协议可以使这两个泛型更加规范和易于扩展。

在 Swift 中，String、Int、Double、Bool，以及无关联值的枚举等都是遵循 Hashable 协议的。Hashable 协议提供的 hashValue 方法用于判断协议对象是否相等。

Hashable 协议同时也是继承 Equatable 协议的，遵循了 Hashable 协议的类或结构通过实现 Equatable 协议确定自定义的类或结构是否相同。在遵循了 Equatable 协议后，协议里的函数就可以对遵循了这个协议的参数进行是否相等的比较了。Swift 的内置类型都支持 Equatable 协议，比如 String、Int 和 Float 等都支持 == 或 != 运算。注意，自定义的类型如果没有遵循 Equatable 协议是没法使用 == 运算或者使用与这个运算相关的方法的，比如数组里的 contain 等。自定义类型一旦实现了 Equatable 协议也就意味着能够使用系统集合里的 API 进行查找，通过重载 >、<、>= 和 <= 等，还能够对自定义的类型在集合里进行排序。

1.2.2 关联类型

在协议里定义的关联类型也可以用泛型来处理。我们定义一个协议，如下所示。

```
protocol HTNState {
    associatedtype StateType
```

```
    func add(_ item: StateType)
}
```

采用非泛型的实现，如下所示。

```
struct states: HTNState {
    typealias StateType = Int
    func add(_ item: Int) {
        //...
    }
}
```

采用泛型遵循协议，如下所示。

```
struct states<T>: HTNState {
    func add(_ item: T) {
        //...
    }
}
```

这样关联类型也能够享受泛型的好处了。

1.2.3 类型擦除

但是在使用关联类型的时候需要注意，当声明一个使用了关联属性的协议对象作为属性时，会先出现 no initializers 的提示，接着会提示 error: protocol 'HTNState' can only be used as a generic constraint because it has Self or associated type requirements 。其意思是 HTNState 协议只能作为泛型约束使用，因为它包含了 Self 或关联类型。

代码如下：

```
class stateDelegate<T> {
    var state: T
    var delegate: HTNState
}
```

那么这个问题该如何处理呢？答案是通过类型擦除来解决，添加一个中间层在代码中，让这个抽象的类型具体化。实际上在 Swift 的标准库里就有对类型擦除的运用，比如 AnySequence 协议。

1.2.4 Where 语句

函数、扩展和关联类型都可以使用 where 语句。where 语句是对泛型在应用时的一种约束。比如：

```
func stateFilter<FromState:HTNState, ToState:HTNState>(_ from:FromState, _ to:ToState) where FromState.StateType == ToState.StateType {
    //...
}
```

这个函数要求它们的 StateType 具有相同类型。

1.2.5 泛型和 Any 类型

泛型和 Any 类型虽然看起来相似，但其实是有区别的。它们的区别在于 Any 类型会避开类型的检查，所以尽量少用或不用。而泛型既灵活又安全。下面举一个例子感受一下两者的区别：

```
func add<T>(_ input: T) -> T {
   //...
   return input;
}

func anyAdd(_ input: Any) -> Any {
   //...
   return input;
}
```

add 函数的 input 参数的类型和函数返回值的类型相同，而 anyAdd 函数的 input 参数的类型和函数返回值的类型不同，这样就会失控，在后续的操作中容易出错。

1.3 集合

1.3.1 基本概念

我们来了解一下集合的基本概念。首先集合是泛型的，比如：

```
let stateArray: Array<String> = ["工作","吃饭","玩游戏","睡觉"]
```

集合需要具备遍历的功能，通过 GeneratorType 协议，可以不关注具体元素类型只要不断地用迭代器调 next 就可以得到全部元素。但是使用迭代器无法进行多次遍历，这时就需要使用 Sequence 协议来解决这个问题。像集合的 forEach、elementsEqual、contains、minElement、maxElement、map、flatMap、filter、reduce 等功能都是使用 Sequence 协议进行多次遍历的。

因为 Sequence 协议无法确定集合里的位置，所以在 Sequence 的基础上增加了 Indexable 协议，Sequence 协议加上 Indexable 协议就是 Collection 协议。有了 Collection 协议就可以确定元素的位置了，包括开始位置和结束位置，这样就能够确定哪些元素是已经访问过的，从而避免多次访问同一个元素，并且通过一个给定的位置直接找到对应位置的元素。

以上过程如下图所示。

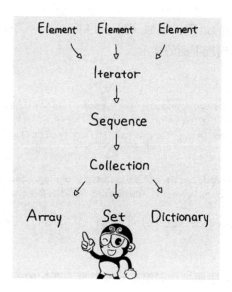

1.3.2 迭代器

Swift 里最简单的迭代器是 AnyIterator，其结构体代码如下：

```
struct AnyIterator<Element>: IteratorProtocol {
  init(_ body: @escaping () -> Element?)
  //...
}
```

AnyIterator 实现了 IteratorProtocol 协议和 Sequence 协议。通过下面的例子我们来看一看如何使用 AnyIterator。

```
class stateItr : IteratorProtocol {
    var num:Int = 1
    func next() -> Int?{
        num += 2
        return num
    }
}

func findNext<I: IteratorProtocol>( elm: I) -> AnyIterator<I.Element> where I.Element == Int
{
    var l = elm
    print("\(l.next() ?? 0)")
    return AnyIterator { l.next() }
}

findNext(elm: findNext(elm: findNext(elm: stateItr())))
```

首先定义一个遵循了 IteratorProtocol 协议并实现了 next 函数的类。然后实现 AnyIterator 的 findNext 方法，通过对这个方法的调用就可以一个接一个地查找符合要求的元素了。

其中有对 where 语句的运用，例如 where I.Element == Int。如果把这句改成 where I.Element == String，则会出现下面的提示。

```
Playground execution failed:

error: MyPlayground.playground:18:37: error: cannot invoke 'findNext(elm:)' with an argument list of type '(elm: stateItr)'
findNext(elm: findNext(elm: findNext(elm: stateItr())))
                                    ^
MyPlayground.playground:11:6: note: candidate requires that the types 'Int' and 'String' be equivalent (requirement specified as 'I.Element' == 'String' [with I = stateItr])
    func findNext<I: IteratorProtocol>( elm: I) -> AnyIterator<I.Element> where I.Element == String
         ^
```

编译器会在代码检查阶段通过代码跟踪发现类型不匹配的安全隐患。对这一功能，我们不得不对 Swift 的设计"点赞"。

1.3.3 Sequence 协议

AnyIterator 只会以单次触发的方式反复计算下一个元素，但是无法重复查找元素或更新已生成的元素。因此还需要有一个新的迭代器和一个子 Sequence 协议。在 Sequence 协议里可以看到如下定义。

```
public protocol Sequence {
    //Element 表示序列元素的类型
    associatedtype Element where Self.Element == Self.Iterator.Element
    //迭代接口类型
    associatedtype Iterator : IteratorProtocol
    //子 Sequence 协议类型
    associatedtype SubSequence
    //返回 Sequence 协议元素的迭代器
    public func makeIterator() -> Self.Iterator
    //...
}
```

重复查找元素可以使用上面 Sequence 协议里的新迭代器。而对于切片这种会更新已生成遵循 Sequence 协议元素的操作，就需要 SubSequence 进行存储和返回了。Sequence 协议支持以迭代的方式来访问元素，可以把 Sequence 对象当成元素的列表，一个接一个地进行访问。此对象还具有以下功能。

- map：对每个元素进行转换并返回新集合对象。
- filter：可设置过滤条件闭包将符合条件的元素返回新集合对象。
- reduce：通过闭包操作元素返回一个值。
- sorted：根据闭包对元素排序，返回排序好的新集合对象。

1.3.4 Collection 协议

Collection 协议继承自 Sequence 协议 和 Indexable 协议，是对 Sequence 协议的进一步完善。最重要的就是使遵循了 Collection 协议的对象和结构体具有下标索引功能，通过查找下标索引能够取到对应的元素。Collection 协议为有限范围，有开始索引和结束索引，这与无限范围的 Sequence 协议是不一样的。有了有限的范围，Collection 协议就可以以 count 属性进行计数了。

除标准库里的 String、Array、Dictionary 和 Set 以外，Data 和 IndexSet 也因为遵循了 Collection 协议而获得了标准库里集合类型的能力。实现 Collection 协议会获得 isEmpty、first 和 count 等方法和属性。

另外还有四种专门用来进行控制特性的集合，分别是 Bidirectional Collection、Random Access Collection、Mutable Collection 和 Range Replaceable Collection，如下图所示。

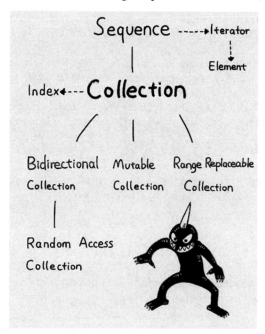

1.3.5 Map

我们先看看 Map 的定义：

```
func map<T>(transform: (Self.Generator.Element) -> T) rethrows -> [T]
```

(Self.Generator.Element) -> T 就是 Map 闭包的定义，Self.Generator.Element 是当前元素的类型。

Map 就是能够通过一个闭包来创建一个数组的映射数组，就像数学中的函数 $y = f(x)$ 的关系一样。下面举一个例子看一看 Map 函数的使用。

```
import UIKit
var cats = ["Tabby", "Snow", "Cycloman"]              ["Tabby","Snow","Cycloman"]
func wearHat(cat: String) -> String {
    return cat + " wear hat"                          (3 times)
}
let catsWithHat = cats.map(wearHat)                   ["Tabby wear hat", "Snow wear hat", "Cycloman wear hat"]
```

1.3.6 flatMap

二维数组经过 flatMap 会降到一维数组，并且过滤掉 nil 值。下面我们看一看 Swift 源代码路径 swift/stdlib/public/core/SequenceAlgorithms.swift.gyb 中 flatMap 的实现。

```
//===----------------------------------------------------------===//
// flatMap()
//===----------------------------------------------------------===//

extension Sequence {
  /// Returns an array containing the concatenated results of calling the
  /// given transformation with each element of this sequence.
  ///
  /// Use this method to receive a single-level collection when your
  /// transformation produces a sequence or collection for each element.
  ///
  /// In this example, note the difference in the result of using 'map' and
  /// 'flatMap' with a transformation that returns an array.
  ///
  ///     let numbers = [1, 2, 3, 4]
  ///
  ///     let mapped = numbers.map { Array(count: $0, repeatedValue: $0) }
  ///     // [[1], [2, 2], [3, 3, 3], [4, 4, 4, 4]]
  ///
  ///     let flatMapped = numbers.flatMap { Array(count: $0, repeatedValue:
  ///       $0) }
  ///     // [1, 2, 2, 3, 3, 3, 4, 4, 4, 4]
  ///
  /// In fact, 's.flatMap(transform)' is equivalent to
  /// 'Array(s.map(transform).joined())'.
  ///
  /// - Parameter transform: A closure that accepts an element of this
  ///   sequence as its argument and returns a sequence or collection.
  /// - Returns: The resulting flattened array.
  ///
  /// - Complexity: O(*m* + *n*), where *m* is the length of this sequence
  ///   and *n* is the length of the result.
  /// - SeeAlso: 'joined()', 'map(_:)'
  public func flatMap<SegmentOfResult : Sequence>(
    _ transform: (${GElement}) throws -> SegmentOfResult
  ) rethrows -> [SegmentOfResult.${GElement}] {
    var result: [SegmentOfResult.${GElement}] = []
    for element in self {
      result.append(contentsOf: try transform(element))
    }
    return result
```

```
    }
  }

  extension Sequence {
    /// Returns an array containing the non-'nil' results of calling the given
    /// transformation with each element of this sequence.
    ///
    /// Use this method to receive an array of nonoptional values when your
    /// transformation produces an optional value.
    ///
    /// In this example, note the difference in the result of using 'map' and
    /// 'flatMap' with a transformation that returns an optional 'Int' value.
    ///
    ///     let possibleNumbers = ["1", "2", "three", "///4///", "5"]
    ///
    ///     let mapped: [Int?] = possibleNumbers.map { str in Int(str) }
    ///     // [1, 2, nil, nil, 5]
    ///
    ///     let flatMapped: [Int] = possibleNumbers.flatMap { str in Int(str) }
    ///     // [1, 2, 5]
    ///
    /// - Parameter transform: A closure that accepts an element of this
    ///   sequence as its argument and returns an optional value.
    /// - Returns: An array of the non-'nil' results of calling 'transform'
    ///   with each element of the sequence.
    ///
    /// - Complexity: O(*m* + *n*), where *m* is the length of this sequence
    ///   and *n* is the length of the result.
    public func flatMap<ElementOfResult>(
      _ transform: (${GElement}) throws -> ElementOfResult?
    ) rethrows -> [ElementOfResult] {
      var result: [ElementOfResult] = []
      for element in self {
        if let newElement = try transform(element) {
          result.append(newElement)
        }
      }
      return result
    }
  }
```

从代码中可以看出 flatMap 的原理是将集合里的所有元素都添加到另外一个新集合里。在上面代码中第二个 extension 里通过 if let 语句过滤集合里解包不成功的元素。我们通过一个例子看一看 flatMap 的使用。注意 Swift 的源代码注释写得很详细,并且有使用的范例可以参考。

```
import UIKit
var cats = ["Tabby", "Snow", "Cycloman"]
var masters = ["Joson", "Curly", "Lana"]
var names = [cats, masters]
func wearHat(name: String) -> String {
    return name + " wear hat"
}
let allWearHat = names.flatMap{$0.map(wearHat)}
print(allWearHat)

["Tabby wear hat", "Snow wear hat", "Cycloman wear hat",
 "Joson wear hat", "Curly wear hat", "Lana wear hat"])
```

1.3.7 Reduce

Reduce 在编程语言语义学里起归约作用，Reduce 也叫"累加器"。下面是其在 Swift 源代码里的实现。

```
//===----------------------------------------------------------===//
// reduce()
//===----------------------------------------------------------===//

extension Sequence {
    /// Returns the result of combining the elements of the sequence using
    /// the given closure.
    ///
    /// Use the 'reduce(_:_:)' method to produce a single value from the
    /// elements of an entire sequence. For example, you can use this method
    /// on an array of numbers to find their sum or product.
    ///
    /// The 'nextPartialResult' closure is called sequentially with an
    /// accumulating value initialized to 'initialResult' and each element
    /// of the sequence. This example shows how to find the sum of an array
    /// of numbers.
    ///
    ///     let numbers = [1, 2, 3, 4]
    ///     let numberSum = numbers.reduce(0, { x, y in
    ///         x + y
    ///     })
    ///     // numberSum == 10
    ///
    /// When 'numbers.reduce(_:_:)' is called, the following steps occur:
    ///
    /// 1. The 'nextPartialResult' closure is called with
    ///    'initialResult'---'0' in this case---and the first element of
    ///    'numbers', returning the sum:'1'.
    /// 2. The closure is called again repeatedly with the previous call's
    ///    return value and each element of the sequence.
    /// 3. When the sequence is exhausted, the last value returned from the
    ///    closure is returned to the caller.
    ///
    /// If the sequence has no elements, 'nextPartialResult' is never executed
    /// and 'initialResult' is the result of the call to 'reduce(_:_:)'.
    ///
```

```
/// - Parameters:
///   - initialResult: The value to use as the initial accumulating value.
///     'initialResult' is passed to 'nextPartialResult' the first time
///     the closure is executed.
///   - nextPartialResult: A closure that combines an accumulating value
///     and an element of the sequence into a new accumulating value, to
///     be used in the next call of the 'nextPartialResult' closure or
///     returned to the caller.
/// - Returns: The final accumulated value. If the sequence has no elements,
///   the result is 'initialResult'.
public func reduce<Result>(
  _ initialResult: Result,
  _ nextPartialResult:
    (_ partialResult: Result, ${GElement}) throws -> Result
) rethrows -> Result {
  var accumulator = initialResult
  for element in self {
    accumulator = try nextPartialResult(accumulator, element)
  }
  return accumulator
}
```

可以看到 Reduce 会通过 initialResult 来记录前面的返回结果和当前元素在闭包里的操作。

1.3.8 Array

Swift 提供了三种数组，分别是 Array、ArraySlice 和 ContiguousArray。其中 ArraySlice 可以在不复制数组时展示数组的局部，生成的局部数组会生成共享缓冲区，缓冲区会延长 ArraySlice 的生命周期，所以 ArraySlice 不适合较大的数组。ContiguousArray 的元素存储在连续的内存区域里。而 Array 在元素不是类或元素遵循了 @objc 协议时也是和 ContiguousArray 一样将元素存储在连续的内存区域里的。

ContiguousArray 和 Array 相比，Array 具有类型检查操作，比如添加元素时的 _isClassOrObjCExistential(Element.self)方法、插入元素时的 isUniquelyReferenced()方法、删除元素时的_makeUniqueAndReserveCapacityIfNotUnique() 方法。

Array 的基本用法如下。

```
//创建数组
var nums = [Int]() //创建空数组
var mArray = nums + [2,3,5] + [5,9]//合并多个有相同类型元素数组的值
var animals: [String] = ["dragon", "cat", "mice", "dog"]

//添加数组
animals.append("bird")
animals += ["ant"]

//获取和改变数组
var firstItem = mArray[0]
animals[0] = "red dragon"
animals[2...4] = ["black dragon", "white dragon"] //使用下标改变多个元素
```

```
animals.insert("chinese dragon", at: 0) //在索引值之前添加元素
let mapleSyrup = animals.remove(at: 0)  //移除数组中的一个元素
let apples = animals.removeLast()  //移除最后一个元素

////数组遍历
for animal in animals {
   print(animal)
}
for (index, animal) in animals.enumerated() {
   print("animal \(String(index + 1)): \(animal)")
}
/*
 animal 1: red dragon
 animal 2: cat
 animal 3: black dragon
 animal 4: white dragon
 */
```

1.3.9 弱引用的 Swift 数组

Swift 里的数组默认使用强引用，但有时候我们希望能够使用弱引用。因此建议使用 NSPointerArray，NSPointerArray 在初始化的时候可以决定是用强引用还是弱引用。

```
let strongArr = NSPointerArray.strongObjects()  // 强引用
let weakArr = NSPointerArray.weakObjects()  // 弱引用
```

Dictionary 要想用弱引用可以使用 NSMapTable，Set 要想用弱引用可以使用 NSHashTable。

1.3.10 Dictionary

Dictionary 的基本用法如下所示。

```
//创建 Dictionary
var strs = [Int: String]()
var colors: [String: String] = ["red": "#e83f45", "yellow": "#ffe651"]
strs[16] = "sixteen"

//updateValue 这个方法会返回更新前的值
if let oldValue = colors.updateValue("#e83f47", forKey: "red") {
   print("The old value for DUB was \(oldValue).")
}

//遍历
for (color, value) in colors {
   print("\(color): \(value)")
}

//map
let newColorValues = colors.map { "hex:\($0.value)" }
print("\(newColorValues)")

//mapValues 返回完整的新 Dictionary
let newColors = colors.mapValues { "hex:\($0)" }
```

```
print("\(newColors)")
```

1.4 协议式编程

Swift 是单继承的,如果要用多继承,则需要使用协议。协议最重要的作用之一就是通过协议设计的 associatedtype 要求使用者遵守指定的泛型约束。

下面我们看一看传统编程的开发模式。

```
class Dragon {

}
class BlackDragon: Dragon{
   func fire() {
      print("fire!!!")
   }
}

class WhiteDragon: Dragon {
   func fire() {
      print("fire!!!")
   }
}

BlackDragon().fire()
WhiteDragon().fire()
```

从上述代码可以看出 BlackDragon 类和 WhiteDragon 类都定义了 fire()。为了避免重复,可以直接在基类里添加 fire()或者通过 extension 来对它们的基类进行扩展。

```
extension Dragon {
   func fire() {
      print("fire!!!")
   }
}
```

此时,添加一个方法让 Dragon 具有"飞"的能力。

```
extension Dragon {
   func fire() {
      print("fire!!!")
   }
   func fly() {
      print("fly~~~")
   }
}
```

如果我们设计出一个新的 Dragon 或者当 Dragon 没有"飞"的能力时,该怎么做呢?因为无法多继承,所以无法拆成两个基类,这样就必然会出现重复代码。但是有了协议,这个问题就好解决了。具体实现如下。

```
protocol DragonFire {}
protocol DragonFly {}
```

```
extension DragonFire {
    func fire() {
        print("fire!!!")
    }
}
extension DragonFly {
    func fly() {
        print("fly~~~")
    }
}

class BlackDragon: DragonFire, DragonFly {}
class WhiteDragon: DragonFire, DragonFly {}
class YellowDragon: DragonFire {}
class PurpleDragon: DragonFire {}

BlackDragon().fire()
WhiteDragon().fire()
BlackDragon().fly()
YellowDragon().fire()
```

可以看到上例中没有了重复代码，结构也清晰了很多并且更容易扩展了。Dragon 的种类和能力的组合也更加清晰。extension 使得协议有了实现默认方法的能力。

关于多继承，Swift 使用了 Trait 方式，其他语言如 C++ 是直接支持多继承的，即一个类会持有多个父类的实例。Java 的多继承只继承多实现，具体实现内容是无法继承的。和 Trait 类似的解决方案是 Mixin。Ruby 就是用的这种元编程思想。

协议可以继承并且可以通过&来聚合，判断一个类是否遵循了一个协议可以使用 is 关键字。

当然，协议还可以作为元素的类型，比如将一个数组里的泛型元素类型指定为协议类型，那么这个数组里的元素类型只要遵循这个协议就可以了。

笔者在一个项目中为了实现将结构化的数据生成符合不同平台规则的代码，做了很多工作。为了能够更好地合并多语言里重复的东西，而将生成不同语言的实现遵循相同的协议。最终效果如下所示。

```
SMNetWorking<H5Editor>()
.requestJSON("https://httpbin.org/get") { (jsonModel) in
    let reStr = H5EditorToFrame<H5EditorObjc>(H5EditorObjc())
.convert(jsonModel)
    print(reStr)
}
```

如果是转成 Swift 的话，那么就把 H5EditorObjc 改成 H5EditorSwift 即可。它们遵循的是 HTNMultilingualismSpecification 协议，其他语言依此类推。如果遇到统一的实现，那么可以建立协议的扩展，然后使用统一函数。

```
extension HTNMultilingualismSpecification {
    //统一函数放这里
}
```

这种设计类似于类簇，比如我们熟悉的 NSString 就是这么设计的。例如，根据初始化的不同，带有 initWith 字符的初始化方法实例化的对象是不同的，但是由于它们都遵循了相同的协议，所以我们在使用的时候没有感觉到差别。

Swift 自带的 JSONDecoder 也是使用协议式编程的典范。decode 函数的定义如下。

```
open func decode<T : Decodable>(_ type: T.Type, from data: Data) throws -> T
```

入参 type 是遵循了统一的 Decodable 协议的，那么就可以按照统一的方法去做处理，在 decode 函数内部实现时，JSONDecoder 会代理给 _JSONDecoder 来实现具体的逻辑。所以在 decode 函数里的具体实现值类型转换的 unbox 函数都是在 _JSONDecoder 的扩展里实现的。unbox 函数会处理数字、字符串、布尔值等基础数据类型。如果有其他层级的结构体也会一层层解下去，_JSONDecoder 的 _JSONDecodingStorage 通过保存基础数据类型最终得到完整的结构体。通过下面的代码我们可以看出支持整个过程的结构是怎么设计的。首先_JSONDecoder 的属性，如下所示。

```
/// The decoder's storage.
fileprivate var storage: _JSONDecodingStorage

/// Options set on the top-level decoder.
fileprivate let options: JSONDecoder._Options

/// The path to the current point in encoding.
fileprivate(set) public var codingPath: [CodingKey]

/// Contextual user-provided information for use during encoding.
public var userInfo: [CodingUserInfoKey : Any] {
   return self.options.userInfo
}
```

下面是初始化代码。

```
/// Initializes 'self' with the given top-level container and options.
fileprivate init(referencing container: Any,
at codingPath: [CodingKey] = [], options: JSONDecoder._Options) {
   self.storage = _JSONDecodingStorage()
   self.storage.push(container: container)
   self.codingPath = codingPath
   self.options = options
}
```

这里可以看到 storage 在初始化时只添加了 container 到顶层，其代码如下。

```
fileprivate mutating func push(container: Any) {
   self.containers.append(container)
}
```

container 在定义的时候是一个 [Any] 数组，因此允许像 container 包含 container、struct 包含 struct 这样的结构出现。

1.5 链式编程

链式编程就是通过点语法来连接不同函数的调用的,每个函数返回的是类对象本身。这种写法使得代码更加简洁,可读性更强。在 iOS 上大家比较熟悉的链式编程运用的是 Masonry 库和 SnapKit 库。

笔者在某项目中,因为使用表达式设置布局属性比如宽、高等,需要兼顾到各种表达式的处理情况,所以设计了一个类似 SnapKit 的可链式调用设置值的结构。先设计一个结构体用来存储一些可变的信息。

```
struct PtEqual {
    var leftId = ""
    var left = WgPt.none
    var leftIdPrefix = ""   //左前缀
    var rightType = PtEqualRightType.pt
    var rightId = ""
    var rightIdPrefix = ""
    var right = WgPt.none
    var rightFloat:Float = 0
    var rightInt:Int = 0
    var rightColor = ""
    var rightText = ""
    var rightString = ""
    var rightSuffix = ""
    
    var equalType = EqualType.normal
}
```

对于这些结构的设置可以在 PtEqualC 类里处理,把每个结构体属性的设置做成各个函数返回类本身即可。效果如下:

```
p.left(.width).leftId(id).leftIdPrefix("self.").rightType(.float)
 .rightFloat(fl.viewPt.padding.left * 2).equalType(.decrease)
```

不过每次设置完后,需要累加到最后返回的字符串里,这样的过程其实也可以封装成一个简单函数,比如 add()。怎么做能够更通用呢? 比如支持不同的累加方法等。

首先可以设计一个累加的 block 属性。

```
typealias MutiClosure = (_ pe: PtEqual) -> String
var accumulatorLineClosure:MutiClosure = {_ in return ""}
```

添加累加字符串和换行标识。

```
var mutiEqualStr = ""              //累加的字符串
var mutiEqualLineMark = "\n"    //换行标识
```

再写一个函数去设置返回是 self 的 block,这样就可以用于链式调用了。

```
//累计设置的字符串
func accumulatorLine(_ closure:@escaping MutiClosure) -> PtEqualC {
    self.accumulatorLineClosure = closure
    return self
}
```

最后添加一个专门用来执行累加动作的函数。

```
//执行累加动作
func add() {
    if filterBl {
        self.mutiEqualStr += accumulatorLineClosure(self.pe) + self.mutiEqualLineMark
    }
    _ = resetFilter()
}
```

我们看一看用起来是什么效果：

```
HTNMt.PtEqualC().accumulatorLine({ (pe) -> String in
    return self.ptEqualToStr(pe: pe)
}).filter({ () -> Bool in
    return vpt.isNormal
}).once({ (p) in
    p.left(.height).rightFloat(fl.viewPt.padding.top * 2).add()
})
```

细心的读者会注意到这里多了两个函数，一个是 filter 函数，另一个是 once 函数。这两个函数里的 block 会将通用逻辑进行封装。当 filter 函数返回值是布尔值时，返回值为 true 的，才会处理后面的 block，以及结构体的属性设置。实现方式如下：

```
//过滤条件
func filter(_ closure: FilterClosure) -> PtEqualC {
    filterBl = closure()
    return self
}
```

以上示例中的 filterBl 是类的一个属性，根据这个属性来决定动作是否继续执行。比如 filterBl 会进行以下判断：

```
func left(_ wp:WgPt) -> PtEqualC {
    filterBl ? self.pe.left = wp : ()
    return self
}
```

once 函数也会进行以下判断：

```
func once(_ closure:(_ pc: PtEqualC) -> Void) -> PtEqualC{
    if filterBl {
        closure(self)
    }
    _ = resetPe()
    _ = resetFilter()
    return self
}
```

同时 once 函数还会重置 filterBl 和设置的结构体，相当于完成了一个完整的设置周期。

有了 filter 函数和 once 函数，再复杂的设置过程，以及逻辑处理都可以很清晰、统一地表达出来了，下面我们分析将复杂 HTML 布局代码映射成原生代码的实现。

```
//UIView *myViewContainer = [UIView new];
lyStr += newEqualStr(vType: .view, id: cId) + "\n"

//属性拼装
lyStr += HTNMt.PtEqualC().accumulatorLine({ (pe) -> String in
    return self.ptEqualToStr(pe: pe)
}).once({ (p) in
    p.left(.top).leftId(cId).end()
    ...
}).once({ (p) in
    p.leftId(cId).left(.left).rightType(.float).rightFloat(0).add()
}).once({ (p) in
    p.leftId(cId).left(.width).rightType(.pt).rightIdPrefix("self.")
    .rightId(id).right(.width).add()
    p.left(.height).right(.height).add()
}).once({ (p) in
    p.left(.width).leftId(id).leftIdPrefix("self.").rightType(.float).
rightFloat(fl.viewPt.padding.left * 2).equalType(.decrease).add()
    p.left(.height).rightFloat(fl.viewPt.padding.top * 2).add()
    ...
}).mutiEqualStr
```

1.6 Swift 内存管理

1.6.1 内存分配

堆

在堆上进行内存分配的时候，需要锁定堆上能够容纳存放对象的空闲块，主要是为了线程安全，我们需要对这些空闲块进行锁定和同步。

堆是完全二叉树，即除最底层节点外都是填满的。底层节点填充是按照从左到右的顺序进行的。Swift 的堆是通过双向链表实现的。由于堆可以 retain 和 release，所以很容易使分配空间不连续。采用链表的目的是希望能够将内存块连起来，在 release 时通过调整链表指针来整合空间。

在 retain 时不可避免地需要遍历堆，找到合适大小的内存块，能优化的也只是记录以前遍历的情况，减少一些遍历。另外，堆是很大的，每次遍历要耗费很长时间，而且 release 为了能够整合空间还需要判断当前内存块的前一块和后一块是否为空闲等。如果空闲，则需要遍历链表查询。所以最终的解决方式是双向链表，只把空闲内存块用指针连起来形成链表。这样在 retain 时可以减少遍历，理论上讲效率可以提高一倍。在 release 时将多余空间插入到堆开始的位置并和插入堆之前的位置进行整合。

即使效率提高了，但是还是比不过栈，所以苹果公司将以前 Objective-C 里一些放在堆里的类型改造成了值类型。

Swift 4 之前的版本的弱引用的做法是在对象的强引用计数为 0 后，看弱引用计数来决定是否保留释放对象所占用的内存。如果弱引用计数不为 0，那么内存还会保留指向的"僵尸"对象直到计数减到 0 时才会被清理掉。为了避免"僵尸"对象长期占用内存，新版的 Swift 引入了可以保存额外信息的 side table，让对象的弱引用先指向其对应的 side table。这样做的好处是，在僵尸对象长期占用内存的情况下，不再需要保留"僵尸"对象，取而代之的是只保留引用计数和指向原对象指针内存占用非常小的 side table。更详细的实现在 Swift 的源代码路径 stdlib/public/SwiftShims/RefCount.h 中。

栈

栈的结构很简单，有 push 和 pop 就可以了。内存上只需要维护栈末端的指针即可。由于它很简单，所以处理一些时效性不强、临时性的代码时，是非常合适的。我们可以把栈看成是一个交换临时数据的内存区域。在多线程上，由于栈是线程独有的，所以也不需要考虑线程的安全问题。

内存对齐

Swift 中也有内存对齐的概念，如下所示。

```
struct DragonFirePosition {
    var x:Int64 //8
    var y:Int32 //4
    //8 + 4
}
struct DragonHomePosition {
    var y:Int32 //4+4
    var x:Int64 //8
    //4 + 4 + 8
}
let firePositionSize = MemoryLayout<DragonFirePosition>.size //12
let homePositionSize = MemoryLayout<DragonHomePosition>.size //16
```

1.6.2 Swift 派发机制

Swift 派发的目的是让 CPU 知道被调用的函数在哪里。Swift 语言支持编译型语言的直接派发、函数表派发、消息机制派发这三种派发机制。下面我们分别对这三种派发机制进行说明。

直接派发

C++ 默认使用的是直接派发，加上 virtual 修饰符可以改成函数表派发。直接派发是最快的，原因是调用指令少，并且可以通过编译器进行比如内联等方式的优化。缺点是由于缺少动态性而不支持继承。

```
struct DragonFirePosition {
    var x:Int64
    var y:Int32
    func land() {}
}

func DragonWillFire(_ position:DragonFirePosition) {
    position.land()
}
```

```
let position = DragonFirePosition(x:342, y:213)
DragonWillFire(position)
```

编译 inline 后，DragonWillFire(DragonFirePosition(x:342, y:213)) 会直接跳到方法实现的地方，结果就变成了 position.land()。

函数表派发

Java 默认就是使用的函数表派发，通过 final 修饰符改成直接派发。函数表派发具有动态性，在 Swift 里函数表叫 Witness Table，在其他语言中函数表叫 Virtual Table。一个类里会用数组来存储里面的函数指针，override 父类的函数会替代以前的函数，子类添加的函数会被加到这个数组里。举一个例子：

```
class Fish {
    func swim() {}
    func eat() {
        //normal eat
    }
}

class FlyingFish: Fish {
    override func eat() {
        //flying fish eat
    }
    func fly() {}
}
```

编译器会给 Fish 类和 FlyingFish 类分别创建 Witness Table。在 Fish 类的函数表里有 swim 函数和 eat 函数，在 FlyingFish 类的函数表里有父类 Fish 的 swim，覆盖了父类的 eat 函数和新增加的 fly 函数。

一个函数被调用时会先去读取对象的函数表，再根据类的地址加上该函数的偏移量得到函数地址，然后跳转到相应的地址上去。从编译后的字节码来看就是两次读取一次跳转，比直接派发慢。

消息机制派发

消息机制派发是在运行时可以改变函数的行为，KVO 和 CoreData 都是对这种机制的运用。Objective-C 默认使用的是消息机制派发，C 语言使用的是直接派发，所以 C 语言派发性能高。Swift 可以通过 dynamic 修饰来支持消息机制派发。

当一个消息被派发，程序运行时就会按照继承关系向上查找被调用的函数。但是这样做的效率不高，所以需要通过缓存来提高效率。这样查找性能就和函数表派发差不多了。

派发的使用场景

根据不同的使用场景，选择不同的派发方式。值类型使用的是直接派发；class 和协议的 extension 使用的是直接派发；class 和协议的初始化声明使用的是函数表派发；class 的@obj extension 使用的是消息机制派发。

派发方式如下所示。

- final：让类里的函数使用直接派发，这样该函数就没有动态性，程序运行时也无法取到这个函数。
- dynamic：可以让类里的函数使用消息机制派发，可以重载 extension 里的函数。

Swift 会在派发上面做优化，比如一个函数没有重载，Swift 就可能会使用直接派发的方式。所以如果属性绑定了 KVO，那么它的 getter 方法和 setter 方法可能会被优化成直接派发，从而导致 KVO 失效。所以记得加上 dynamic 的修饰来保证 KVO 有效。我们期待 Swift 能够在之后的版本迭代中针对派发做更多的优化。

1.6.3 基本数据类型内存管理

通过 MemoryLayout，我们来看看基本数据类型的内存占用有多大。

```
MemoryLayout<Int>.size        //8
MemoryLayout<Int16>.size      //2
MemoryLayout<Bool>.size       //1
MemoryLayout<Float>.size      //4
MemoryLayout<Double>.size     //8
```

1.6.4 struct 内存管理

由于 struct 在编译中就能够确定内存占用空间的大小，所以程序运行时不需要额外的内存空间，程序运行过程中函数调用就是直接传地址。

下面我们再看一看 struct 的 MemoryLayout。

```
struct DragonFirePosition {
   var x:Int64 //8
   var y:Int32 //4
   //8 + 4
   func land() {}
}

MemoryLayout<DragonFirePosition>.size        //12
MemoryLayout<DragonFirePosition>.alignment   //8
MemoryLayout<DragonFirePosition>.stride      //16
```

可以看出 alignment 是按照 8 字节来对齐的。这里 struct 用到了字节对齐，实际占用大小为 16 字节。

如果把 var x:Int64 改成可选类型，那么会增加 4 字节。不过就这个示例来说其实际大小还是 16 字节。这也是因为内存对齐的原因。

1.6.5 class 内存管理

class 本身是在 Stack 上分配的，在 Heap 上还需要保存 class 的 type 信息。type 信息里有一个函数表，class 的函数在派发时会按照 type 信息里的函数表进行派发。当子类需要继承父类时，子类只要在自己的 type 信息里记录自己的信息即可。

1.6.6 协议类型内存管理

协议类型内存管理使用的是 Existential Container 内存模型。我们先看下面的例子：

```
protocol DragonFire {}
extension DragonFire {
   func fire() {
      print("fire!!!")
   }
}

struct YellowDragon: DragonFire {
   let eyes = "blue"
   let teeth = 48
}

let classSize = MemoryLayout<YellowDragon>.size //32
let protocolSize = MemoryLayout<DragonFire>.size //40
```

上面例子中的结构体即遵循了 DragonFire 协议，而且结构体的内容还比 DragonFire 协议多，但 DragonFire 协议的内存比结构体大，这是为什么呢？原因是 Existential Container 的前三个 word 是用 Value buffer 来存储 inline 的值的。第四个 word 是 Value Witness Table，即用来存储值的各种操作，比如 allocate、copy、destruct 和 deallocate 等。第五个 word 是 Protocol Witness Table，即用来存储协议的函数。

1.6.7 泛型的内存管理

泛型采用的是和 Existential Container 原理类似的内存管理。Value Witness Table 和 Protocol Witness Table 作为隐形的参数传递到泛型方法里。不过经过编译器的层层 inline 优化，最终类型会被推导出来，也就不再需要 Existential Container 这一套方法了。

1.7 JSON 数据的处理

做项目只要是涉及服务器端接口，就无法避免和 JSON 数据打交道。对于来自网络的 JSON 数据的处理，可以使用苹果系统自带的字符串转模型类 JSONDecoder。这个类在 Swift 4 的 Foundation 模块里。我们可以在 Swift 源代码目录 swift/stdlib/public/SDK/Foundation/JSONEncoder.swift 看到苹果系统对这个类的实现。

其他对 JSON 处理的库还有 SwiftyJSON。

1.7.1 使用 JSONDecoder

下面是使用 JSONDecoder 的案例。

```
struct GroceryProduct: Codable {
   var name: String
   var points: Int
   var description: String?
```

```
}
let json = """
{
   "name": "Durian",
   "points": 600,
   "description": "A fruit with a distinctive scent."
}
""".data(using: .utf8)!

let decoder = JSONDecoder()
let product = try decoder.decode(GroceryProduct.self, from: json)

print(product.name) // Prints "Durian"
```

要注意的是 GroceryProduct 结构体需要遵循 Codable 协议。因为 JSONDecoder 的实例对象的 decode 函数的参数是一个遵循 Decodable 协议的结构体。Codable 协议是 Encodable 和 Decodable 两个协议的组合，写法如下所示。

```
public typealias Codable = Decodable & Encodable
```

当然 JSON 数据的结构不会都这么简单，如果遇到以下嵌套情况，则可以通过在 struct 里再套一个 struct 来完成。

```
let json = """
{
   "name": "Durian",
   "points": 600,
   "ability": {
      "mathematics": "excellent",
      "physics": "bad",
      "chemistry": "fine"
   },
   "description": "A fruit with a distinctive scent."
}
""".data(using: .utf8)!
```

修改过的 struct 如下所示。

```
struct GroceryProduct: Codable {
   var name: String
   var points: Int
   var ability: Ability
   var description: String?
   
   struct Ability: Codable {
      var mathematics: String
      var physics: String
      var chemistry: String
   }
}
```

可以观察到在 Ability 里数学、物理、化学的评价只有优、良、差几种，所以很适合用枚举表示。Swift 的枚举对于字符串关联类型的枚举也有很好的支持,只要声明关联值类型是 String 就行了,修改后的代码如下所示。

```
struct GroceryProduct: Codable {
    var name: String
    var points: Int
    var ability: Ability
    var description: String?

    struct Ability: Codable {
        var mathematics: Appraise
        var physics: Appraise
        var chemistry: Appraise
    }

    enum Appraise: String, Codable {
        case excellent, fine, bad
    }
}
```

API 返回的结果会有一个不可控的因素,是什么呢?那就是有的键值有时会返回有时不会返回,那么 struct 要怎么兼容呢?

好在 Swift 原生支持 optional,只要在属性后加个问号即可。比如 points 有时会返回,有时不会返回,那么就可以像下面这样写:

```
struct GroceryProduct: Codable {
    var name: String
    var points: Int? //可能会用不到
}
```

1.7.2 CodingKey 协议

接口还会有一些其他不可控因素。例如,在现实生活中不同团队、不同公司的风格是不同的。而 JSONDecoder 考虑到了不同代码风格的问题,其允许通过映射的方式实现对不同代码风格的兼容。下面我们看一看如何用:

```
let json = """
{
    "nick_name": "Tom",
    "points": 600,
}
""".data(using: .utf8)!
```

上面示例中的 nick_name 我们希望处理成 Swift 的风格,那么可以使用一个遵循 CodingKey 协议的枚举来做映射。

```
struct GroceryProduct: Codable {
    var nickName: String
    var points: Int

    enum CodingKeys : String, CodingKey{
```

```
        case nickName = "nick_name"
        case points
    }
}
```

上面这个方法是通用方法，可以处理各种风格的命名，比如喜欢简写，那么 nickName 就可以写成 nName。Codingkey 协议默认的实现，实际上已经能够解决现实环境的大部分问题了。如果有些自定义的字段映射要处理的话，那么可以通过覆盖默认 Codingkey 协议的方式来完成。通过获取 encoder 的 container 对象进行自定义操作。

1.7.3 JSONDecoder 的 keyDecodingStrategy 属性

JSONDecoder 里还有一个专门的属性——keyDecodingStrategy。这个属性的数据类型是枚举，其中的一个枚举成员是 convertFromSnakeCase，可以按照 snake case 的策略来进行值的转换。这个策略是核心功能内置的，就不需要我们额外写代码处理了。可以不用实现 CodingKey 协议，只需要在 JSONDecoder 实例中设置 keyDecodingStrategy 属性即可。

```
let decoder = JSONDecoder()
decoder.keyDecodingStrategy = .convertFromSnakeCase
```

keyDecodingStrategy 属性是在 Swift 4.1 中加上的，所以 Swift 4.1 之前的处理值还是用 CodingKey 协议来处理的。

那么如何通过设置 keyDecodingStrategy 属性做到直接使用策略来做值转换的呢？

实现值转换功能的代码就在 Swift 源代码目录 swift/stdlib/public/ SDK/Foundation/下的 JSONEncoder.swift 文件里。

我们先看一看 keyDecodingStrategy 属性的定义：

```
/// The strategy to use for decoding keys. Defaults to '.useDefaultKeys'.
open var keyDecodingStrategy: KeyDecodingStrategy = .useDefaultKeys
```

keyDecodingStrategy 属性是一个 KeyDecodingStrategy 枚举类型，默认的枚举值是 .userDefaultKeys。定义如下所示。

```
public enum KeyDecodingStrategy {
    /// Use the keys specified by each type. This is the default strategy.
    case useDefaultKeys

    /// Convert from "snake_case_keys" to "camelCaseKeys" before attempting
    /// to match a key with the one specified by each type.
    ///
    /// The conversion to upper case uses 'Locale.system', also known as
    /// the ICU "root" locale. This means the result is consistent regardless
    /// of the current user's locale and language preferences.
    ///
    /// Converting from snake case to camel case:
    /// 1. Capitalizes the word starting after each '_'
    /// 2. Removes all '_'
    /// 3. Preserves starting and ending '_' (as these are often used to
    /// indicate private variables or other metadata).
```

```
/// For example, 'one_two_three' becomes 'oneTwoThree'.
/// '_one_two_three_' becomes '_oneTwoThree_'.
///
/// - Note: Using a key decoding strategy has a nominal performance cost,
/// as each string key has to be inspected for the '_' character.
/// case convertFromSnakeCase

/// Provide a custom conversion from the key in the encoded JSON to the
/// keys specified by the decoded types.
/// The full path to the current decoding position is provided for context
/// (in case you need to locate this key within the payload). The returned
/// key is used in place of the last component in the coding path before
/// decoding. If the result of the conversion is a duplicate key,
/// then only one value will be present in the container for the type
/// to decode from.
case custom((_ codingPath: [CodingKey]) -> CodingKey)

fileprivate static func _convertFromSnakeCase(_ stringKey: String) -> String {
    ...
}
```

convertFromSnakeCase 就是用来进行值转换的函数，注释部分描述了整个过程，首先函数会把下画线符号后面的字母转成大写字母，然后除最前面和最后的下画线以外，还要移除所有其他的下画线。比如 _nick_name 就会转换成 _nickName。而这些转换操作是在枚举定义的静态方法 _convertFromSnakeCase 里完成的，代码如下所示。

```
fileprivate static func _convertFromSnakeCase(_ stringKey: String) -> String {
    guard !stringKey.isEmpty else { return stringKey }

    // Find the first non-underscore character
    guard let firstNonUnderscore = stringKey.index(where: { $0 != "_" }) else {
        // Reached the end without finding an _
        return stringKey
    }

    // Find the last non-underscore character
    var lastNonUnderscore = stringKey.index(before: stringKey.endIndex)
    while lastNonUnderscore > firstNonUnderscore && stringKey[lastNonUnderscore] == "_" {
        stringKey.formIndex(before: &lastNonUnderscore);
    }

    let keyRange = firstNonUnderscore...lastNonUnderscore
    let leadingUnderscoreRange = stringKey.startIndex..<firstNonUnderscore
    let trailingUnderscoreRange = stringKey.index(after: lastNonUnderscore)..<stringKey.endIndex

    var components = stringKey[keyRange].split(separator: "_")
    let joinedString : String
    if components.count == 1 {
```

```
        // No underscores in key, leave the word as is - maybe already
        // camel cased
        joinedString = String(stringKey[keyRange])
    } else {
        joinedString = ([components[0].lowercased()] + components[1...]
.map { $0.capitalized }).joined()
    }

    // Do a cheap isEmpty check before creating and appending potentially
    //empty strings
    let result : String
    if (leadingUnderscoreRange.isEmpty
      && trailingUnderscoreRange.isEmpty) {
        result = joinedString
    } else if (!leadingUnderscoreRange.isEmpty
      && !trailingUnderscoreRange.isEmpty) {
        // Both leading and trailing underscores
        result = String(stringKey[leadingUnderscoreRange]) + joinedString
            + String(stringKey[trailingUnderscoreRange])
    } else if (!leadingUnderscoreRange.isEmpty) {
        // Just leading
        result = String(stringKey[leadingUnderscoreRange]) + joinedString
    } else {
        // Just trailing
        result = joinedString + String(stringKey[trailingUnderscoreRange])
    }
    return result
}
```

上面这段代码的逻辑处理不是很复杂,功能也不多,但是仍有很多值得学习的地方,首先我们可以看它是如何处理边界条件的。其中两个边界条件都是用 guard 语句来处理的,如下所示。

```
guard !stringKey.isEmpty else { return stringKey }

// Find the first non-underscore character
guard let firstNonUnderscore = stringKey.index(where: { $0 != "_" })
else {
    // Reached the end without finding an _
    return stringKey
}
```

第一个是判断空,第二个是通过 String 的 public func index(where predicate: (Character) throws -> Bool) rethrows -> String.Index?检查字符串 stringKey 里是否包含了下画线。如果没有包含,就直接返回原 String 值。index(where predicate: (Character) throws -> Bool) rethrows -> String.Index? 函数的参数就是一个自定义返回布尔值的 block,返回 true,则立刻返回不再继续遍历了。可见苹果公司对于性能一点也不浪费。

返回的 index 值还可以得到第一个下画线后面的第一个非下画线的字符。因此,不需要把最前面和最后面的下画线转驼峰了。但是前面的下画线和后面的下画线个数不确定,所以需要得到从前面下画线到后面下画线的范围。

取得范围后，接下来要怎么做呢？从下面的示例中可以看到对 String 的 public func formIndex(before i: inout String.Index) 函数的应用。这里的参数定义为 inout，作用是能够在函数里对这个参数不用通过返回的方式直接修改生效。这个函数的作用就是移动字符串 stringKey 的 index。before 是往前移动，after 是往后移动。

```
// Find the last non-underscore character
var lastNonUnderscore = stringKey.index(before: stringKey.endIndex)
while lastNonUnderscore > firstNonUnderscore
    && stringKey[lastNonUnderscore] == "_" {
    stringKey.formIndex(before: &lastNonUnderscore);
}
```

上面的代码就是先找到整个字符串的最后的 index，然后开始从后往前找，找到不是下画线的字符时就跳出 while，同时还要满足不超过 lastNonUnderscore 范围的要求。

现在我们可以考虑这样一个问题，为什么在做前面的判断时不用 public func formIndex(after i: inout String.Index)函数呢？因为 after 代表从前往后移动遍历，所以也可以达到找到第一个不是下画线的字符就停止的效果。

这一方式可以说是一箭双雕，既解决了边界问题又满足了需求，优化了代码并且减少了代码量。笔者作为面试官在 iOS 面试过程中，通常会出算法题，很多人以为只要有了解决思路或者能写出简单的处理代码就可以了，甚至有些人以为用中文一条一条写出思路就行了。其实不然，考察分为两种，一种是考察智商，即考察候选人是否能想出更多的方法解决问题，针对的是那些在面谈过程中实力不错的人，更多是为了判断面试人是否具有创造力，属于拔尖的考法。另外一种是考察实际项目能力，考察边界条件的处理、逻辑的严谨性，以及对代码优化的处理，相关题的解法和逻辑会比较简单。

_convertFromSnakeCase 枚举的静态函数会在创建 container 的时候调用，具体使用的函数是 _JSONKeyedDecodingContainer。在 _JSONKeyedDecodingContainer 的初始化方法里会判断 decoder.options.keyDecodingStrategy 的枚举值是否满足 convertFromSnakeCase。如果满足，就调用_convertFromSnakeCase 枚举的静态函数。调用的时候要注意的是转换成驼峰后的 key 可能会和已有 key 的名称重名。那么就需要看选择哪个值了，苹果公司选择的是第一个值。其实现方式如下所示。

```
self.container = Dictionary(container.map {
    key, value in
    (JSONDecoder.KeyDecodingStrategy._convertFromSnakeCase(key), value)
}, uniquingKeysWith: { (first, _) in first })
```

其中 Dictionary 的初始化函数，如下所示。

```
public init<S>(_ keysAndValues: S, uniquingKeysWith combine:
(Dictionary.Value, Dictionary.Value) throws -> Dictionary.Value)
rethrows where S : Sequence, S.Element == (Key, Value)
```

init 函数就是专门用来处理重复 key 的问题的。如果要选择最后一个 key 的值，那么用 init 函数也会很容易实现，如下所示。

```
let pairsWithDuplicateKeys = [("a", 1), ("b", 2), ("a", 3), ("b", 4)]
```

```
let firstValues = Dictionary(pairsWithDuplicateKeys,
                    uniquingKeysWith: { (first, _) in first })
// ["b": 2, "a": 1]
let lastValues = Dictionary(pairsWithDuplicateKeys,
                    uniquingKeysWith: { (_, last) in last })
// ["b": 4, "a": 3]
```

1.7.4　枚举定义 block

KeyEncodingStrategy 可以自定义 CodingKey，如下所示。

```
case custom((_ codingPath: [CodingKey]) -> CodingKey)
```

在 container 初始化时，函数会调用 block 来进行 key 的转换。同样如果转换后出现重复 key，函数也会和 convertFromSnakeCase 一样选择第一个值。这里可以看到 Swift 里的枚举能够定义 block，方便自定义 key 转换处理规则，这样就可以完全抛弃以前那种为了实现 CodingKey 进行 key 转换的方式了。

1.7.5　inout

在 Swift 源代码中可以看到 formIndex(before i: inout Index)函数是如何实现的。

在源代码里找到 formIndex(before i: inout Index)函数时，发现还有一个同名函数，却没有 inout 定义：

```
public func index(before i: Index) -> Index {
    if i.value == i.extent.lowerBound {
        // Move to the next range
        if i.rangeIndex == 0 {
            // We have no more to go
            return Index(value: i.value, extent: i.extent, rangeIndex: i.rangeIndex, rangeCount: i.rangeCount)
        } else {
            let rangeIndex = i.rangeIndex - 1
            let rangeCount = i.rangeCount
            let extent = _range(at: rangeIndex)
            let value = extent.upperBound - 1
            return Index(value: value, extent: extent, rangeIndex: rangeIndex, rangeCount: rangeCount)
        }
    } else {
        // Move to the previous value in this range
        return Index(value: i.value - 1, extent: i.extent, rangeIndex: i.rangeIndex, rangeCount: i.rangeCount)
    }
}

public func formIndex(before i: inout Index) {
    if i.value == i.extent.lowerBound {
        // Move to the next range
        if i.rangeIndex == 0 {
            // We have no more to go
```

```
        } else {
            i.rangeIndex -= 1
            i.extent = _range(at: i.rangeIndex)
            i.value = i.extent.upperBound - 1
        }
    } else {
        // Move to the previous value in this range
        i.value -= 1
    }
}
```

对 index 和 formIndex 这两个函数的实现，最直观的感受就是 inout 的 formIndex 函数少了三个 return。其好处就是值类型参数 i 可以以引用方式传递，不需要 var 和 let 来修饰。

另外，inout 还有一个好处在上面的 formIndex 函数里没有体现出来，那就是可以方便地对多个值类型数据进行修改而不需要一一指明返回。

1.8 网络请求

说到网络请求，在 Objective-C 里基本用的是 AFNetworking，而在 Swift 里用的则是 Alamofire 库。笔者在 Swift 1.0 之前的 beta 版本中就注意到了 Alamofire 库。那时还是 Mattt Thompson 一个人在写，文件也只有一个。如今其功能已经完善了很多。笔者在做 HTN 项目时，项目对于网络请求的需求不是很大。有需要的时候，就是使用 URLSession 简单地实现网络请求，直接拉下接口下发的 JSON 数据即可。

结合前面解析 JSON 的方法，网络请求的代码如下所示。

```
struct WebJSON:Codable {
    var name:String
    var node:String
    var version: Int?
}
let session = URLSession.shared
let request:URLRequest = NSURLRequest.init(url: URL(string:
"http://www.starming.com/api.php?get=testjson")!) as URLRequest
let task = session.dataTask(with: request) { (data, res, error) in
    if (error == nil) {
        let decoder = JSONDecoder()
        do {
            print("解析 JSON 成功")
            let jsonModel = try decoder.decode(WebJSON.self, from: data!)
            print(jsonModel)
        } catch {
            print("解析 JSON 失败")
        }
    }
}
```

以上代码可以成功请求到 JSON 数据，然后转换成对应的结构数据。但如果还有另外几处也要进行网络请求，那么这种方式就不合适了。我们先看一看 Alamofire 库是如何操作的，如下所示。

```
Alamofire.request("https://httpbin.org/get").responseData { response in
    if let data = response.data {
        let decoder = JSONDecoder()
        do {
            print("解析 JSON 成功")
            let jsonModel = try decoder.decode(H5Editor.self, from: data)
        } catch {
            print("解析 JSON 失败")
        }
    }
}
```

Alamofire 库有 responseJSON 方法，但解析完是字典，用的时候需要做很多容错判断，很不方便。所以还是要使用 JSONDecoder 或者其他第三方库。不过 Alamofire 库的写法已经做了一些简化，并且还实现了更多的功能，这些之后再说，现在的主要任务是简化调用。于是笔者开始动手修改先前的实现。首先创建一个网络类，然后简化 request 方法，建一个 block 方便请求完成后对网络返回数据的处理。最后使用泛型，支持不同 struct 的数据统一返回。写完后，给这个网络类起个名字叫"SMNetWorking"，其实现如下。

```
open class SMNetWorking<T:Codable> {
    open let session:URLSession
    
    typealias CompletionJSONClosure = (_ data:T) -> Void
    var completionJSONClosure:CompletionJSONClosure = {_ in }
    
    public init() {
        self.session = URLSession.shared
    }
    
    //JSON 的请求
    func requestJSON(_ url: SMURLNetWorking,
                    doneClosure:@escaping CompletionJSONClosure
                    ) {
        self.completionJSONClosure = doneClosure
        let request:URLRequest = NSURLRequest.init(url: url.asURL()) as URLRequest
        let task = self.session.dataTask(with: request) {
            (data, res, error) in
            if (error == nil) {
                let decoder = JSONDecoder()
                do {
                    print("解析 JSON 成功")
                    let jsonModel = try decoder.decode(T.self, from: data!)
                    self.completionJSONClosure(jsonModel)
                } catch {
                    print("解析 JSON 失败")
                }
                
            }
        }
        task.resume()
```

```
    }
}

/*----------Protocol----------*/
protocol SMURLNetWorking {
    func asURL() -> URL
}

/*----------Extension----------*/
extension String: SMURLNetWorking {
    public func asURL() -> URL {
        guard let url = URL(string:self) else {
            return URL(string:"http:www.starming.com")!
        }
        return url
    }
}
```

修改后，调用就简单多了，如下所示。

```
SMNetWorking<WModel>()
.requestJSON("https://httpbin.org/get") { (jsonModel) in
    print(jsonModel)
}
```

特别当请求不同的接口，返回不同结构时，本地定义了很多 model 结构体，那么请求时只需要指明不同的 model 结构体，block 就能够直接返回对应的值。

默认请求都按照 GET 方法请求，在实际项目中会用到其他方法比如 POST 等，使用 Alamofire 库的方法如下所示。

```
/// HTTP method definitions.
///
/// See https://tools.ietf.org/html/rfc7231#section-4.3
public enum HTTPMethod: String {
    case options = "OPTIONS"
    case get     = "GET"
    case head    = "HEAD"
    case post    = "POST"
    case put     = "PUT"
    case patch   = "PATCH"
    case delete  = "DELETE"
    case trace   = "TRACE"
    case connect = "CONNECT"
}
```

先定义一个枚举，依据的标准已在代码注释里说明。使用效果如下所示。

```
Alamofire.request("https://httpbin.org/get") // method defaults to '.get'
Alamofire.request("https://httpbin.org/post", method: .post)
Alamofire.request("https://httpbin.org/put", method: .put)
Alamofire.request("https://httpbin.org/delete", method: .delete)
```

可以看到在 request 方法里有一个可选参数，设置完后，会给 NSURLRequest 的

httpMethod 的可选属性附上设置的值。如下所示。

```
public init(url: URLConvertible, method: HTTPMethod
, headers: HTTPHeaders? = nil) throws {
    let url = try url.asURL()

    self.init(url: url)

    httpMethod = method.rawValue

    if let headers = headers {
        for (headerField, headerValue) in headers {
            setValue(headerValue, forHTTPHeaderField: headerField)
        }
    }
}
```

接下来，在 SMNetWorking 类里也加上了这个功能，先定义一个枚举：

```
enum HTTPMethod: String {
    case GET,OPTIONS,HEAD,POST,PUT,PATCH,DELETE,TRACE,CONNECT
}
```

利用枚举的字符串协议特性，将枚举的字符串值和枚举名对应上，减少赋值代码从而简化枚举定义。

NSURLRequest 提供的可选设置项较多，如果把这些设置都做成一个个可配参数，那么后期维护会非常麻烦。建议使用链式方法来完成设置。先从设置 httpMethod 开始，如下所示。

```
//链式方法
//httpMethod 的设置
func httpMethod(_ md:HTTPMethod) -> SMNetWorking {
    self.op.httpMethod = md
    return self
}
```

上面代码中的 op 是一个结构体，专门用来存放可选项的值。

使用起来也很方便：

```
SMNetWorking<WModel>().method(.POST).requestJSON("https://httpbin.org/get")
```

有了这样一个结构的设计，之后进行扩展时会非常方便。不过目前的功能只要满足基本需求即可，其他需要完善的地方，可以先提供一个接口到外部去设置。我们先建一个 block 把 URLRequest 转到外部的接口进行设置。

完成后的使用效果如下：

```
SMNetWorking<WModel>().method(.POST).configRequest { (request) in
    //设置 request
}.requestJSON("https://httpbin.org/get")
```

就刚才提到的请求参数来说，Alamofire 定义了一个 ParameterEncoding 协议，在该协议里规定了统一处理请求参数的方法 func encode(_ urlRequest:URLRequestConvertible,with parameters: Parameters?) throws -> URLRequest。这样就可以对多种情况做一样的返回处理了。可以看到 URL、JSON 和 PropertyList 都遵循了 ParameterEncoding 协议。encode 函数实现了 ParameterEncoding 协议。encode 函数在实现中会将 URL、JSON、PropertyList 赋值给 httpBody 属性。以 URLEncoding 为例，我们来看一看具体实现：

```
public func encode(_ urlRequest: URLRequestConvertible, with parameters:
Parameters?) throws -> URLRequest {
    var urlRequest = try urlRequest.asURLRequest()

    guard let parameters = parameters else { return urlRequest }

    if let method = HTTPMethod(rawValue: urlRequest.httpMethod ?? "GET"),
encodesParametersInURL(with: method) {
        guard let url = urlRequest.url else {
            throw AFError.parameterEncodingFailed(reason: .missingURL)
        }

        if var urlComponents = URLComponents(url: url, resolvingAgainstBaseURL:
false), !parameters.isEmpty {
            let percentEncodedQuery = (urlComponents.percentEncodedQuery.map { $0
+ "&" } ?? "") + query(parameters)
            urlComponents.percentEncodedQuery = percentEncodedQuery
            urlRequest.url = urlComponents.url
        }
    } else {
        if urlRequest.value(forHTTPHeaderField: "Content-Type") == nil {
            urlRequest.setValue("application/x-www-form-urlencoded;
charset=utf-8", forHTTPHeaderField: "Content-Type")
        }

        urlRequest.httpBody = query(parameters).data(using: .utf8,
allowLossyConversion: false)
    }

    return urlRequest
}
```

1.9 自动布局 SnapKit 库分析

笔者认为 SnapKit 库的框架设计和封装做得很好，本节将围绕"给谁做约束？如何设置约束？设置完后如何处理？"这三个问题分析 SnapKit 库是怎么做的。

首先，我们看一看 SnapKit 框架的整体结构图，如下图所示。对它们有个大概的印象。

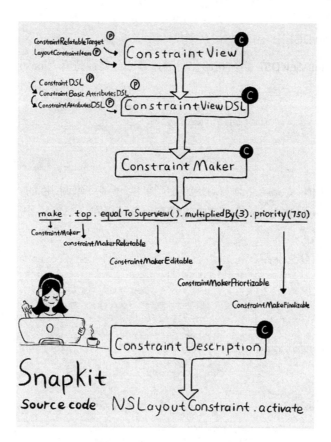

1.9.1 给谁做约束

ConstraintView

ConstraintView 实际上在 iOS 里就是 UIView，在 macOS 上就是 NSView，如下所示。

```
#if os(iOS) || os(tvOS)
   public typealias ConstraintView = UIView
#else
   public typealias ConstraintView = NSView
#endif
```

对 ConstraintView 做扩展，在扩展里定义一个 snp 属性，如下所示。

```
public extension ConstraintView {
    public var snp: ConstraintViewDSL {
        return ConstraintViewDSL(view: self)
    }
}
```

snp 属性就是结构体 ConstraintViewDSL。下面我们来看一看结构体 ConstraintViewDSL 的作用。

ConstraintViewDSL

结构体 ConstraintViewDSL 会在初始化时通过 view 属性持有 ConstraintView，如下所示。

```
internal let view: ConstraintView

internal init(view: ConstraintView) {
    self.view = view

}
```

同时，结构体 ConstraintViewDSL 还提供了必须调用的 makeConstraints、contentHuggingHorizontalPriority 等函数。这样我们就可以在 UIView 中直接调用这些函数来进行视图的约束设置了。

```
public func makeConstraints(_ closure: (_ make: ConstraintMaker) -> Void) {
    ConstraintMaker.makeConstraints(item: self.view, closure: closure)
}

public var contentHuggingHorizontalPriority: Float {
    get {
        return self.view.contentHuggingPriority(for: .horizontal)
        .rawValue
    }
    set {
        self.view.setContentHuggingPriority(LayoutPriority(rawValue: newValue),
for: .horizontal)
    }
}
//还有 remakeConstraints、contentCompressionResistanceHorizontalPriority 等就不一一列出了
...
```

结构体 ConstraintViewDSL 继承自 ConstraintAttributesDSL。

ConstraintAttributesDSL

ConstraintAttributesDSL 是一个协议，继承自协议 ConstraintBasicAttributesDSL。为什么要多一层呢？因为 ConstraintAttributesDSL 里定义了 iOS 8 系统出现的新的属性，例如 lastBaseline、firstBaseline、leftMargin 等。而 ConstraintBasicAttributesDSL 里定义的是一开始就有的那些属性，例如 left、top、centerX、size 等。

1.9.2 如何设置约束

ConstraintMaker

ConstraintMaker 是设置约束的入口，makeConstraints 函数的一个闭包参数可以在外部设置 ConstraintMaker 的 left、right、top 等属性以描述约束。这些属性的 getter 方法会返回 ConstraintMakerExtendable 实例。

ConstraintMaker 的构造函数,如下所示。

```
internal init(item: LayoutConstraintItem) {
    self.item = item
    self.item.prepare()
}
```

LayoutConstraintItem 会通过扩展 ConstraintLayoutGuide 和 ConstraintView 来起到约束 item 类型的作用。prepare 函数的作用,如下所示。

```
internal func prepare() {
    if let view = self as? ConstraintView {
        view.translatesAutoresizingMaskIntoConstraints = false
    }
}
```

在 prepare 函数中统一禁用了 AutoresizeMask。闭包参数设置属性的 getter 方法,如下所示。

```
public var left: ConstraintMakerExtendable {
    return self.makeExtendableWithAttributes(.left)
}

internal func makeExtendableWithAttributes(_ attributes:
    ConstraintAttributes) -> ConstraintMakerExtendable {
    let description = ConstraintDescription(item: self.item, attributes:
attributes)
    self.descriptions.append(description)
    return ConstraintMakerExtendable(description)
}
```

ConstraintMaker 包含了一个 ConstraintDescription 数组,里面会记录用户设置的各个属性,然后返回 ConstraintMakerExtendable。

这里的 ConstraintAttributes 是结构体,ConstraintAttributes 结构体遵从的是 OptionSet 协议。为什么不用枚举呢? 因为枚举每次只能选择一个选项。而在 Swift 里的枚举是没法将多个枚举选项组成一个值的,比如 ConstraintAttributes 里的 edges、size 和 center 等就是由多个枚举选项组合而成的。而 OptionSet 结构体使用了高效的位域来进行表示。另外,OptionSet 继承自 ExpressibleByArrayLiteral,因此可以使用数组字面量法来生成选项的集合。下面看一看 ConstraintAttributes 是如何定义的。

```
internal static var none: ConstraintAttributes { return self.init(0) }
internal static var left: ConstraintAttributes { return self.init(1) }
internal static var top: ConstraintAttributes { return self.init(2) }
internal static var right: ConstraintAttributes { return self.init(4) }
internal static var bottom: ConstraintAttributes { return self.init(8) }
internal static var leading: ConstraintAttributes { return self.init(16) }
internal static var trailing: ConstraintAttributes { return self.init(32) }
internal static var width: ConstraintAttributes { return self.init(64) }
internal static var height: ConstraintAttributes { return self.init(128) }
internal static var centerX: ConstraintAttributes { return self.init(256) }
internal static var centerY: ConstraintAttributes { return self.init(512) }
```

```
//就不一一列出来了
...
// 组合
internal static var edges: ConstraintAttributes { return self.init(15) }
internal static var size: ConstraintAttributes { return self.init(192) }
internal static var center: ConstraintAttributes { return self.init(768) }

@available(iOS 8.0, *)
internal static var margins: ConstraintAttributes { return self.init(61440) }
//还有一些，先不列了
...
```

可以看到 size 属性就是 width(64) + height(128) = size(192)。

ConstraintAttributes 重载了 +、+=、-= 和 == 运算符，代码如下所示。

```
internal func + (left: ConstraintAttributes, right: ConstraintAttributes)
  -> ConstraintAttributes {
    return left.union(right)
}

internal func +=(left: inout ConstraintAttributes, right:
  ConstraintAttributes) {
    left.formUnion(right)
}

internal func -=(left: inout ConstraintAttributes, right:
  ConstraintAttributes) {
    left.subtract(right)
}

internal func ==(left: ConstraintAttributes, right: ConstraintAttributes)
  -> Bool {
    return left.rawValue == right.rawValue
}
```

以上重载很适合对自定义的结构体进行简化符号的操作。

如果想要自定义运算符的话，则需要进行声明，让编译器知道这是运算符。比如自定义一个运算符🐱，代码如下所示。

```
struct A {
    var v:Int = 0
}
infix operator 🐱
func 🐱(left: A, right: A) -> Bool {
    return left.v + 1 == right.v
}
```

infix 为中间运算符，prefix 为前置运算符，postfix 为后置运算符。自定义运算符可由 /、=、-、+、*、%、<、>、!、&、|、^、。、~ 等组成，或者使用一些特殊的符号，比如上例中的🐱，以及 ⊕、∉ 等。

自定义运算符还能指定优先级分组，示例如下。

```
infix operator ⊆ : CPrecedence
precedencegroup CPrecedence {
    associativity: left
    higherThan: AdditionPrecedence
    lowerThan: MultiplicationPrecedence
}
```

常用类型对应的 group，如下所示。

```
// "Exponentiative"
infix operator <<  : BitwiseShiftPrecedence
infix operator &<< : BitwiseShiftPrecedence
infix operator >>  : BitwiseShiftPrecedence
infix operator &>> : BitwiseShiftPrecedence

// "Multiplicative"
infix operator *  : MultiplicationPrecedence
infix operator &* : MultiplicationPrecedence
infix operator /  : MultiplicationPrecedence
infix operator %  : MultiplicationPrecedence
infix operator &  : MultiplicationPrecedence

// "Additive"
infix operator +  : AdditionPrecedence
infix operator &+ : AdditionPrecedence
infix operator -  : AdditionPrecedence
infix operator &- : AdditionPrecedence
infix operator |  : AdditionPrecedence
infix operator ^  : AdditionPrecedence

// FIXME: is this the right precedence level for "..." ?
infix operator ...  : RangeFormationPrecedence
infix operator ..<  : RangeFormationPrecedence
```

完整的运算符的定义和优先级关系在 Swift 源代码的 swift/stdlib/public/core/Policy.swift 文件里有详细介绍。哪些运算符优先级高或哪些运算符优先级低在这个文件里一目了然。

ConstraintMakerExtendable

ConstraintMakerExtendable 继承自 ConstraintMakerRelatable，它可以实现多个属性的链式设置，包括 left、right、top 等，以产生一个 ConstraintMakerRelatable 类型的实例。left 属性的 getter 定义如下：

```
public var left: ConstraintMakerExtendable {
    self.description.attributes += .left
    return self
}
```

从中可以看到通过重载的运算符 += 能够将 .left 加到 ConstraintAttributes 里。

ConstraintMakerRelatable

ConstraintMakerRelatable 的作用是指定视图间的约束关系，比如常用的 equalTo 函数。

通过 relatedTo 函数返回 ConstraintMakerEditable 类型的实例。

```swift
    @discardableResult
    public func equalTo(_ other: ConstraintRelatableTarget, _ file: String = #file, _ line: UInt = #line) -> ConstraintMakerEditable {
        return self.relatedTo(other, relation: .equal, file: file, line: line)
    }
    internal func relatedTo(_ other: ConstraintRelatableTarget, relation: ConstraintRelation, file: String, line: UInt) -> ConstraintMakerEditable {
        let related: ConstraintItem
        let constant: ConstraintConstantTarget

        if let other = other as? ConstraintItem {
            guard other.attributes == ConstraintAttributes.none ||
                other.attributes.layoutAttributes.count <= 1 ||
                other.attributes.layoutAttributes == self.description
                .attributes.layoutAttributes || other.attributes == .edges &&
                self.description.attributes == .margins ||
                other.attributes == .margins && self.description
                .attributes == .edges else {
                    fatalError("Cannot constraint to multiple non identical attributes. (\(file), \(line))");
            }

            related = other
            constant = 0.0
        } else if let other = other as? ConstraintView {
            related = ConstraintItem(target: other, attributes: ConstraintAttributes.none)
            constant = 0.0
        } else if let other = other as? ConstraintConstantTarget {
            related = ConstraintItem(target: nil, attributes: ConstraintAttributes.none)
            constant = other
        } else if #available(iOS 9.0, OSX 10.11, *), let other = other as? ConstraintLayoutGuide {
            related = ConstraintItem(target: other, attributes: ConstraintAttributes.none)
            constant = 0.0
        } else {
            fatalError("Invalid constraint. (\(file), \(line))")
        }

        let editable = ConstraintMakerEditable(self.description)
        editable.description.sourceLocation = (file, line)
        editable.description.relation = relation
        editable.description.related = related
        editable.description.constant = constant
        return editable
    }
```

上例中 ConstraintRelatableTarget 是约束视图的实例，equalTo 函数中能传的参数类型比较多，ConstraintRelatableTarget 的作用是设置可支持类型，对 ConstraintRelatableTarget 进行扩展，添加

更多的可支持类型。ConstraintPriorityTarget、ConstraintInsetTarget、ConstraintOffsetTarget 也有类似的作用。它们另外的作用就是将 Float、Double、Int 和 UInt 这几种类型都转成 CGFloat。我们以 ConstraintInsetTarget 为例，看其是如何实现的，代码如下所示。

```
extension ConstraintInsetTarget {

    internal var constraintInsetTargetValue: ConstraintInsets {
        if let amount = self as? ConstraintInsets {
            return amount
        } else if let amount = self as? Float {
            return ConstraintInsets(top: CGFloat(amount), left: CGFloat(amount), bottom: CGFloat(amount), right: CGFloat(amount))
        } else if let amount = self as? Double {
            return ConstraintInsets(top: CGFloat(amount), left: CGFloat(amount), bottom: CGFloat(amount), right: CGFloat(amount))
        } else if let amount = self as? CGFloat {
            return ConstraintInsets(top: amount, left: amount, bottom: amount, right: amount)
        } else if let amount = self as? Int {
            return ConstraintInsets(top: CGFloat(amount), left: CGFloat(amount), bottom: CGFloat(amount), right: CGFloat(amount))
        } else if let amount = self as? UInt {
            return ConstraintInsets(top: CGFloat(amount), left: CGFloat(amount), bottom: CGFloat(amount), right: CGFloat(amount))
        } else {
            return ConstraintInsets(top: 0, left: 0, bottom: 0, right: 0)
        }
    }

}
```

ConstraintMakerEditable

ConstraintMakerEditable 继承自 ConstraintMakerPriortizable，主要作用是提供 offset、inset、multipliedBy、dividedBy 这些用来设置约束的函数。

```
public class ConstraintMakerEditable: ConstraintMakerPriortizable {

    @discardableResult
    public func multipliedBy(_ amount: ConstraintMultiplierTarget) -> ConstraintMakerEditable {
        self.description.multiplier = amount
        return self
    }

    @discardableResult
    public func dividedBy(_ amount: ConstraintMultiplierTarget) -> ConstraintMakerEditable {
        return self.multipliedBy(1.0 / amount
            .constraintMultiplierTargetValue)
    }

    @discardableResult
```

```
        public func offset(_ amount: ConstraintOffsetTarget) ->
ConstraintMakerEditable {
            self.description.constant = amount.constraintOffsetTargetValue
            return self
        }

        @discardableResult
        public func inset(_ amount: ConstraintInsetTarget) -> ConstraintMakerEditable {
            self.description.constant = amount.constraintInsetTargetValue
            return self
        }

    }
```

ConstraintMakerPriortizable

ConstraintMakerPriortizable 继承自 ConstraintMakerFinalizable，主要作用是设置优先级，返回 ConstraintMakerFinalizable 类型的实例。

ConstraintMakerFinalizable

ConstraintMakerFinalizable 是类型为 ConstraintDescription 的类，用来描述一个完整的约束，有了 ConstraintDescription 就可以进行后面的处理了。ConstraintMakerFinalizable 对约束的描述是完整的，ConstraintMakerFinalizable 是一个描述类，用于描述一条具体的约束关系，以及包括 ConstraintAttributes 在内的各种与约束有关的元素。一个 ConstraintDescription 实例就可以提供与一种约束有关的所有内容，如下所示。

```
    public class ConstraintDescription {

        internal let item: LayoutConstraintItem
        internal var attributes: ConstraintAttributes
        internal var relation: ConstraintRelation? = nil
        internal var sourceLocation: (String, UInt)? = nil
        internal var label: String? = nil
        internal var related: ConstraintItem? = nil
        internal var multiplier: ConstraintMultiplierTarget = 1.0
        internal var constant: ConstraintConstantTarget = 0.0
        internal var priority: ConstraintPriorityTarget = 1000.0
        internal lazy var constraint: Constraint? = {
            guard let relation = self.relation,
                let related = self.related,
                let sourceLocation = self.sourceLocation else {
              return nil
            }
            let from = ConstraintItem(target: self.item, attributes: self.attributes)

            return Constraint(
              from: from,
              to: related,
              relation: relation,
              sourceLocation: sourceLocation,
```

```
            label: self.label,
            multiplier: self.multiplier,
            constant: self.constant,
            priority: self.priority
        )
    }()
    
    // MARK: Initialization
    
    internal init(item: LayoutConstraintItem, attributes: ConstraintAttributes) {
        self.item = item
        self.attributes = attributes
    }
    
}
```

1.9.3 设置完约束后如何处理

通过 makeConstraints，我们来看看 ConstraintMaker 是如何在外部通过一个闭包写约束关系的，如下所示。

```
internal static func makeConstraints(item: LayoutConstraintItem, closure: (_ make: ConstraintMaker) -> Void) {
    let maker = ConstraintMaker(item: item)
    closure(maker)
    var constraints: [Constraint] = []
    for description in maker.descriptions {
        guard let constraint = description.constraint else {
            continue
        }
        constraints.append(constraint)
    }
    for constraint in constraints {
        constraint.activateIfNeeded(updatingExisting: false)
    }
}
```

在闭包为 maker 的 ConstraintMaker 实例中写入信息，遍历 maker 的 description 之后（我们之前说一个约束最终得到一个 self.description，但往往会有多条约束，所以 ConstraintMakerFinalizable 里面的 self.description，在 ConstraintMaker 里被一个数组维护），我们得到了 constraint 数组。跟进 constraint 数组里的 activateIfNeeded 函数，分析约束是怎么写出来的，如下所示。

```
internal func activateIfNeeded(updatingExisting: Bool = false) {
    guard let item = self.from.layoutConstraintItem else {
        print("WARNING: SnapKit failed to get from item from constraint. Activate will be a no-op.")
        return
    }
    let layoutConstraints = self.layoutConstraints
```

```swift
        if updatingExisting {
            var existingLayoutConstraints: [LayoutConstraint] = []
            for constraint in item.constraints {
                existingLayoutConstraints += constraint.layoutConstraints
            }

            for layoutConstraint in layoutConstraints {
                let existingLayoutConstraint = existingLayoutConstraints
                    .first { $0 == layoutConstraint }
                guard let updateLayoutConstraint = existingLayoutConstraint else {
                    fatalError("Updated constraint could not find existing matching constraint to update: \(layoutConstraint)")
                }

                let updateLayoutAttribute = (updateLayoutConstraint
                    .secondAttribute == .notAnAttribute) ? updateLayoutConstraint
                    .firstAttribute : updateLayoutConstraint.secondAttribute
                updateLayoutConstraint.constant = self.constant
                    .constraintConstantTargetValueFor(layoutAttribute:
                    updateLayoutAttribute)
            }
        } else {
            NSLayoutConstraint.activate(layoutConstraints)
            item.add(constraints: [self])
        }
    }
```

1.10 用 Swift 实现一个简单的语言转译器

1.10.1 转译器简介

在后面章节里我们会讲很多跟编译器相关的内容，为了给后面的内容打基础，本节会讲解如何实现一个 Scheme 函数调用代码转译成 C 语言代码的转译器。了解语言的转译器实现方式有助于理解编译器的原理，在编译器里会不断地将语言转换降维，直到转成机器能够理解的程度。

这里使用 Scheme 语言做例子的原因是笔者觉得当下很多的语言都是功能的堆砌，而 Scheme 只需要很少的表达式构造规则，并且组合方式没有限制，就可以支持现在各种主要的编程范式，是高效、灵活、实用的语言。如果用 ES6 的 JavaScript，那么其复杂的表达式结构设计会让新人感到疑惑，并且用过长的代码做示例，效果不好。

Scheme 是 Lisp 的一种"方言"，另一种"方言"是 Common Lisp。这两个"方言"最大的区别在于规范的不同，Scheme 只有几十页规范，而 Common Lisp 由于自带了巨大的函数库，所以规范达到了上千页。Lisp 诞生于 1975 年，是一门很"古老"的语言，其"古老"程度仅次于 Fortran。Lisp 的代码里到处都是体现递归嵌套的小括号。按照递归理论，所有计算函数都可以归为几种基本的递归函数的不同组合。

Lisp 和它的"方言"们本质上的区别是对作用域处理的不同。这也是为什么 Lisp 会被"方言"取代的根本原因。Lisp 采用的是动态作用域，而 Scheme、Common Lisp，甚至如今流行的 Python、JavaScript 都使用的是静态作用域。动态作用域是在运行时基于调用栈才确定的，函数在不同的地方调用作用域，不同的调用栈有不同的作用域。静态作用域只在声明定义的地方确定，函数和闭包采用同样的数据结构，由函数的定义和当前的运行环境组成。运行环境由一个变量和值的映射表表示。这样在找变量值时，只要在对应的环境映射表里去找值即可，而不像动态作用域那样在调用时要去全局映射表里去找离自己最近的一个值，这种找值的方式太动态、太难查错。

以下为 Scheme 的函数调用示例：

```
(multiply (add 1.4 3))
```

对应的 C 语言代码如下：

```
multiply(add(1.4, 3))
```

接下来我们就来写一个程序完成这个转译过程。这个程序不会包含全部语法，只是希望通过简单的例子让大家能够快速了解转译器的核心思想。

1.10.2　词法分析器

虽然以上示例中的转译器很小，但是还是会包含编译器的词法分析、语法分析、遍历、转换和代码生成等部分。我们先从词法分析开始学习。整个过程就是把代码切成一个又一个的 token，这个 token 可大可小。在示例中我们将其按照数字、符号、分隔符和字符来分就可以了。

首先创建一个 token 的结构体用来记录 token。

```
public struct JToken {
    var type = ""
    var value = ""
}
```

我们希望输入(multiply (add 1.4 3))后，输出以下 token：

```
[HTN.JToken(type: "paren", value: "("),
HTN.JToken(type: "char", value: "add"),
HTN.JToken(type: "int", value: "2"),
HTN.JToken(type: "paren", value: "("),
HTN.JToken(type: "char", value: "subtract"),
HTN.JToken(type: "float", value: "4.4"),
HTN.JToken(type: "int", value: "2"),
HTN.JToken(type: "paren", value: ")"),
HTN.JToken(type: "paren", value: ")")]
```

解析代码需要一个字符一个字符地分析，所以我们先设计一个变量记录当前 String 的 index，如下所示。

```
private var _index: String.Index
```

使用 formIndex 函数移动访问下一个字符。因为 formIndex 函数可以利用 inout 特性，所以比 Index 更加简洁。

```
func advanceIndex() {
    _input.formIndex(after: &_index)
}
```

按照以下代码编写函数得到当前字符,返回值使用 optional 类型作为 while 的中断条件。

```
var currentChar: Character? {
    return _index < _input.endIndex ? _input[_index] : nil
}
```

到达最后一个字符后返回 nil 值。此时在 while 循环里不断调用 advanceIndex currentChar 就会不断返回下一个字符。

```
while let aChar = currentChar {
    advanceIndex()
}
```

这个简单的 Scheme 代码里只涉及三个符号,左括号、右括号和空格。所以我们把它们放到一个条件里进行处理。遇到这些符号时,如果是空格,就直接跳过;如果是括号,就添加到 tokens 堆栈集合里记录一下即可。

```
let symbols = ["(",")"," "]
if symbols.contains(s) {
    if s == " " {
        //空格
        advanceIndex()
        continue
    }
    //特殊符号
    tokens.append(JToken(type: "paren", value: s))
    advanceIndex()
    continue
}
```

以上代码的处理对于一门完整的语言来讲其实并不是最完善的。比如在词法分析中,一般空格和括号会生成单独的 token。但是当空格和括号在一个字符串或一条注释中时,就不能生成单独的 token 了,而需要和字符串或注释里的其他字符整体生成一个 token。所以在词法分析过程中,就需要增加一个状态来记录分析的状态,不同状态生成不同类型的 token。由于本章节的核心知识在于讲解完整的转译过程,所以完善的处理会在后面的章节里再讲到。

剩下的工作就是处理字符和数字。在 else 条件里增加一个 while,代码如下所示。

```
} else {
    var word = ""
    while let sChar = currentChar {
        let str = sChar.description
        if symbols.contains(str) {
            break
        }
        word.append(str)
        advanceIndex()
        continue
```

```
   }
   //开始把连续字符进行 token 存储
   if word.count > 0 {
      var tkType = "char"
      if word.isFloat() {
         tkType = "float"
      }
      if word.isInt() {
         tkType = "int"
      }
      tokens.append(JToken(type: tkType, value: word))
   }
   continue
} // end if
```

如果字符是我们定义的左括号、右括号和空格这三个符号，while 循环就会 break，否则就将符合条件的字符拼接上。当 while 循环 break 后，函数再判断类型是字符、整数还是浮点数。最后根据函数判断出的类型生成一个 token 并添加到 tokens 里。这里对拼接好字符串进行类型判断的函数在自定义的 String 的扩展里。代码如下所示。

```
extension String {
   // 判断是否是整数
   func isInt() -> Bool {
      let scan:Scanner = Scanner(string: self)
      var val:Int = 0
      return scan.scanInt(&val) && scan.isAtEnd
   }
   // 判断是否是 Float
   func isFloat() -> Bool {
      let scan:Scanner = Scanner(string: self)
      var val:Float = 0
      return scan.scanFloat(&val) && scan.isAtEnd
   }
}
```

完整的 tokenizer 代码如下所示。

```
public class JTokenizer {
   private var _input: String
   private var _index: String.Index

   public init(_ input: String) {
      _input = input.filterAnnotationBlock()
      _index = _input.startIndex
   }

   public func tokenizer() -> [JToken] {
      var tokens = [JToken]()
      while let aChar = currentChar {
         let s = aChar.description
         let symbols = ["(",")"," "]
         if symbols.contains(s) {
            if s == " " {
```

```
                    //空格
                    advanceIndex()
                    continue
                }
                //特殊符号
                tokens.append(JToken(type: "paren", value: s))
                advanceIndex()
                continue
            } else {
                var word = ""
                while let sChar = currentChar {
                    let str = sChar.description
                    if symbols.contains(str) {
                        break
                    }
                    word.append(str)
                    advanceIndex()
                    continue
                }
                //开始把连续字符进行 token 存储
                if word.count > 0 {
                    var tkType = "char"
                    if word.isFloat() {
                        tkType = "float"
                    }
                    if word.isInt() {
                        tkType = "int"
                    }
                    tokens.append(JToken(type: tkType, value: word))
                }
                continue
            } // end if
        } // end while
        
        return tokens
    }
    
    //parser tool
    var currentChar: Character? {
        return _index < _input.endIndex ? _input[_index] : nil
    }
    func advanceIndex() {
        _input.formIndex(after: &_index)
    }
}
```

接下来输入 Scheme 代码看一看解析的效果。

```
let a = JTokenizer("(multiply (add 1.4 3))").tokenizer()
print("\(a)")
```

结果如下所示。

```
[ListToC.JToken(type: "paren", value: "("),
```

```
ListToC.JToken(type: "char", value: "multiply"),
ListToC.JToken(type: "paren", value: "("),
ListToC.JToken(type: "char", value: "add"),
ListToC.JToken(type: "float", value: "1.4"),
ListToC.JToken(type: "int", value: "3"),
ListToC.JToken(type: "paren", value: ")"),
ListToC.JToken(type: "paren", value: ")")]
```

1.10.3 语法分析器

语法分析的目的是将数组转换成树状结构，使得各个树上的各个节点之间形成关系。所以需要根据函数调用表达式设计一个节点的结构体。结构体里需要有能够关联不同参数节点的参数堆栈节点、表明节点类型的类型枚举节点、记录节点值的值节点和值类型。首先定义值类型和节点类型，如下所示。

```
// 值类型
public enum JNumberType {
    case int,float
}
// 节点类型
public enum JNodeType {
    case None
    case NumberLiteral
    case CallExpression
}
```

节点的结构体可以按照以下方式进行设计。

```
public protocol JNodeBase {
    var type: JNodeType {get}
    var name: String {get}
    var params: [JNode] {get}
}
public protocol JNodeNumberLiteral {
    var numberType: JNumberType {get}
    var intValue: Int {get}
    var floatValue: Float {get}
}
// struct
public struct JNode:JNodeBase,JNodeNumberLiteral {
    public var type = JNodeType.None
    public var name = ""
    public var params = [JNode]()
    public var numberType = JNumberType.int
    public var intValue:Int = 0
    public var floatValue:Float = 0
}
```

示例中的词法分析过程非常简单。由于只有函数调用和参数设置的环节，所以只需要在参数为另一个函数调用时进行递归处理即可。

```
// 解析类
public class JParser {
```

```swift
    private var _tokens: [JToken]
    private var _current: Int

    public init(_ input:String) {
        _tokens = JTokenizer(input).tokenizer()
        _current = 0
    }
    public func parser() -> [JNode] {
        _current = 0
        var nodeTree = [JNode]()
        while _current < _tokens.count {
            nodeTree.append(walk())
        }
        _current = 0  //用完重置
        return nodeTree
    }

    private func walk() -> JNode {
        var tk = _tokens[_current]
        var jNode = JNode()
        //检查是不是数字类型节点
        if tk.type == "int" || tk.type == "float" {
            _current += 1
            jNode.type = .NumberLiteral
            if tk.type == "int", let intV = Int(tk.value) {
                jNode.intValue = intV
                jNode.numberType = .int
            }
            if tk.type == "float", let floatV = Float(tk.value) {
                jNode.floatValue = floatV
                jNode.numberType = .float
            }
            return jNode

        }
        //检查是否是 CallExpression 类型
        if tk.type == "paren" && tk.value == "(" {
            //跳过符号
            _current += 1
            tk = _tokens[_current]

            jNode.type = .CallExpression
            jNode.name = tk.value
            _current += 1
            while tk.type != "paren" ||
            (tk.type == "paren" && tk.value != ")") {
                //递归下降
                jNode.params.append(walk())
                tk = _tokens[_current]
            }
            //跳到下一个
            _current += 1
            return jNode
        }
```

```
        _current += 1
        return jNode
    }
}
```

walk 函数为递归函数,先处理满足数字类型的节点,对这种节点的处理是直接返回该节点。函数调用节点的起始状态是通过左小括号来判断的,碰到小括号类型的 token 会先创建一个 CallExpression 的节点,然后开启 while 循环去找满足右小括号的 token。满足前,函数会递归并将返回的节点添加到当前节点的参数集合 params 里。代码如下所示。

```
// --------- 打印 AST, 方便调试 ---------
private func astPrintable(_ tree:[JNode]) {
    for aNode in tree {
        recDesNode(aNode, level: 0)
    }
}
private func recDesNode(_ node:JNode, level:Int) {
    let nodeTypeStr = node.type
    var preSpace = ""
    for _ in 0...level {
        if level > 0 {
            preSpace += "  "
        }
    }
    var dataStr = ""
    switch node.type {
    case .NumberLiteral:
        var numberStr = ""
        if node.numberType == .float {
            numberStr = "\(node.floatValue)"
        }
        if node.numberType == .int {
            numberStr = "\(node.intValue)"
        }
        dataStr = "number type is \(node.numberType) number is \(numberStr)."
    case .CallExpression:
        dataStr = "expression is \(node.type)(\(node.name))"
    case .None:
        dataStr = ""
    }
    print("\(preSpace) \(nodeTypeStr) \(dataStr)")

    if node.params.count > 0 {
        for aNode in node.params {
            recDesNode(aNode, level: level + 1)
        }
    }
}
```

打印出的节点树如下所示。

```
CallExpression expression is CallExpression(multiply)
    CallExpression expression is CallExpression(add)
```

```
NumberLiteral number type is float number is 1.4
NumberLiteral number type is int number is 3
```

1.10.4 遍历器

设计一个 traverser 类作为遍历器，用一个键值结构作为记录，将不同类型节点的回调处理闭包作为入参。

```
public func traverser(visitor:[String:VisitorClosure])
```

在遍历到对应的节点时，通过对应类型的 key 去执行对应的闭包。

```
public func traverser(visitor:[String:VisitorClosure]) {

    func traverseChildNode(childrens:[JNode], parent:JNode) {
        for child in childrens {
            traverseNode(node: child, parent: parent)
        }
    }

    func traverseNode(node:JNode, parent:JNode) {
        //执行外部传入的 closure
        if visitor.keys.contains(node.type.rawValue) {
            if let closure:VisitorClosure = visitor[node.type.rawValue] {
                closure(node,parent)
            }
        }
        //看是否有子节点需要继续遍历
        if node.params.count > 0 {
            traverseChildNode(childrens: node.params, parent: node)
        }
    }
    let rootNode = JNode()
    rootNode.type = .Root
    traverseChildNode(childrens: _ast, parent: rootNode)
}
```

1.10.5 转换器

转换器要做的事情就是，生成的 Scheme 节点树通过前面的遍历器将需要处理的闭包传进 traverser 函数里，以构建 C 语言函数调用的节点树。

设计 Transformer 作为执行该任务的类，在构造函数里执行 JTraverser 的 traverser 函数。将 traverser 函数需要的回调闭包在构造函数里写好。我们先看一看 NumberLiteral 这个类型节点的回调闭包，由于这个节点是没有子节点的，所以不需要将 currentParent 设置为 NumberLiteral 类型节点。

此时 NumberLiteral 类型节点的父节点有两种，一种是 ExpressionStatement 类型节点，另一种是 CallExpression 类型节点。这两种类型节点都需要将当前的 NumberLiteral 类型节点添加到父节点的 arguments 里。实现如下：

```
let numberLiteralClosure:VisitorClosure = { (node,parent) in
```

```
    if currentParent.type == .ExpressionStatement {
       currentParent.expressions[0].arguments.append(node)
    }
    if currentParent.type == .CallExpression {
       currentParent.arguments.append(node)
    }
}
```

接下来是对 CallExpression 类型节点的处理。它分为两种情况,一种是父节点也是 CallExpression 类型节点;另一种是父节点不是 CallExpression 类型节点,则需要判断是否是 Root 类型的根节点。

如果不是 CallExpression 类型节点,就需要生成一个新的 ExpressionStatement 类型节点。在父节点是 Root 类型的情况下,将其添加到新的 AST 的根下,然后把 currentParent 设为新生成的 ExpressionStatement 类型节点。

如果是 CallExpression 类型节点,那么父节点在以上示例里就是 ExpressionStatement 类型节点且这个 CallExpression 类型节点在以上示例里就一定为参数。我们只需将其添加到 ExpressionStatement 的 expressions 的 arguments 里即可,具体代码实现如下:

```
let callExpressionClosure:VisitorClosure = { (node,parent) in
    let exp = JNode()
    exp.type = .CallExpression
    
    let callee = JNodeCallee()
    callee.type = .Identifier
    callee.name = node.name
    exp.callee = callee
    
    if parent.type != .CallExpression {
       let exps = JNode()
       exps.type = .ExpressionStatement
       exps.expressions.append(exp)
       if parent.type == .Root {
          self.ast.append(exps)
       }
       currentParent = exps
    } else {
       currentParent.expressions[0].arguments.append(exp)
       currentParent = exp
    }
}
```

1.10.6 代码生成器

代码生成器依赖于之前生成的 C 语言函数调用的节点树。实质上就是通过递归调用将不同的节点类型生成对应的符号和关键字,最后将所有字符组成一个长的字符串。代码如下所示。

```
public init(_ input:String) {
    let ast = JTransformer(input).ast
    for aNode in ast {
       code.append(recGeneratorCode(aNode))
    }
```

```
        print("The code generated:")
        print(code)
    }

    public func recGeneratorCode(_ node:JNode) -> String {
        var code = ""
        if node.type == .ExpressionStatement {
            for aExp in node.expressions {
                code.append(recGeneratorCode(aExp))
            }
        }
        if node.type == .CallExpression {
            code.append(node.callee.name)
            code.append("(")
            if node.arguments.count > 0 {
                for (index,arg) in node.arguments.enumerated() {
                    code.append(recGeneratorCode(arg))
                    if index != node.arguments.count - 1 {
                        code.append(", ")
                    }
                }
            }
            code.append(")")
        }
        if node.type == .Identifier {
            code.append(node.name)
        }
        if node.type == .NumberLiteral {
            switch node.numberType {
            case .float:
                code.append(String(node.floatValue))
            case .int:
                code.append(String(node.intValue))
            }
        }

        return code
    }
```

至此，我们就完成了从一个语言代码转换到另一个语言代码的任务，其过程和结果如下所示。

```
    Input code is:
    (multiply (add 1.4 3))
    Tokens:
    [ListToC.JToken(type: "paren", value: "("), ListToC.JToken(type: "char", value:
"multiply"), ListToC.JToken(type: "paren", value: "("), ListToC.JToken(type: "char",
value: "add"), ListToC.JToken(type: "float", value: "1.4"), ListToC.JToken(type: "int",
value: "3"), ListToC.JToken(type: "paren", value: ")"), ListToC.JToken(type: "paren",
value: ")")]
    Before transform AST:
    CallExpression expression is CallExpression(multiply)
        CallExpression expression is CallExpression(add)
            NumberLiteral number type is float number is 1.4
```

```
            NumberLiteral number type is int number is 3
After transform AST:
ExpressionStatement
    CallExpression expression is CallExpression(multiply)
        CallExpression expression is CallExpression(add)
            NumberLiteral number type is float number is 1.4
            NumberLiteral number type is int number is 3
The code generated:
multiply(add(1.4, 3))
```

1.10.7 Scheme 的其他特性

大家可以将 Scheme 的其他特性转译成 C 语言代码，如表 1-1 所示。

表 1-1

Scheme	C 语言
(< a x b)	((a < x) && (x < b))
(foo a b)	foo(a,b)
(define (square a) (* a a))	int square(int a) { return (a * a); }

1.10.8 Babel

前面我们实现了一个简单的转译器。如果想实现对一个语言完整的转译该怎样做才是最佳实践呢？Babel 是专门处理 JavaScript 的转译工作的，它的使用非常广泛，它能够将 ES6 的代码转成 ES5 的代码。

Babel 的工作流程和前面我们实现的转译器一模一样。babel-parser（曾被称为 "babylon"）通过词法分析器和语法分析器把 ES5 的代码生成抽象语法树(AST，Abstract Syntax Tree)。babel-traverse 遍历器会以节点的 type 作为 key，函数闭包 Visitor 作为值在 AST 递归遍历的过程中，使用转换器通过回调当前节点和父节点封装成一个 path 给闭包使用，从而生成新的 AST。最后使用代码生成器 babel-generator 将生成的新 AST 转化成 ES5 代码。

babel-parser 里包含了词法分析器和语法分析器，派生自 Acorn。其生成的 AST 的节点设计标准来自 ESTree，包含了 ES5、ES2105，等等。

使用在线工具 astexplorer 能够在线查看多种语言的 AST，包括 JavaScript、PHP、Lua，甚至 HTML 和 CSS 的 AST。我们以 JavaScript 代码为例，看一看 Acorn 生成的 AST 是什么样子的。在 astexplorer 里输入：

```
Math.abs((0.1 + 0.2) - 0.3)
```

生成的 AST JSON 数据如下：

```
{
  "type": "Program",
  "start": 0,
  "end": 27,
  "body": [
    {
      "type": "ExpressionStatement",
      "start": 0,
      "end": 27,
      "expression": {
        "type": "CallExpression",
        "start": 0,
        "end": 27,
        "callee": {
          "type": "MemberExpression",
          "start": 0,
          "end": 8,
          "object": {
            "type": "Identifier",
            "start": 0,
            "end": 4,
            "name": "Math"
          },
          "property": {
            "type": "Identifier",
            "start": 5,
            "end": 8,
            "name": "abs"
          },
          "computed": false
        },
        "arguments": [
          {
            "type": "BinaryExpression",
            "start": 9,
            "end": 26,
            "left": {
              "type": "BinaryExpression",
              "start": 10,
              "end": 19,
              "left": {
                "type": "Literal",
                "start": 10,
                "end": 13,
                "value": 0.1,
                "raw": "0.1"
              },
              "operator": "+",
              "right": {
                "type": "Literal",
                "start": 16,
                "end": 19,
                "value": 0.2,
                "raw": "0.2"
```

```
                }
              },
              "operator": "-",
              "right": {
                "type": "Literal",
                "start": 23,
                "end": 26,
                "value": 0.3,
                "raw": "0.3"
              }
            }
          ]
        }
      }
    ],
    "sourceType": "module"
}
```

一个函数调用表达式的 AST 就这样生成了。Acorn 库支持完整的 JavaScript 语法，任何符合 ES 标准的 JavaScript 都能够生成符合语法标准的 AST。大家可以手动输入不同的 JavaScript 代码观察和学习 ES 标准。

Babel 能够将代码转译的结果输出为 JSON，并且 JSON 里包含了转译后的 AST 和代码。因此我们可以把 Babel 当作一个接口来使用，比如可以使用 Swift 对 Babel 接口返回的 JSON 进行词法分析和语法分析，而不必使用 Babel 的插件来分析。接下来我们看一看如何让 Swift 能够调用 Babel 生成的 AST JSON 数据。

首先需要安装 Babel 的工具链。在要使用的工程目录下新建一个 package.json，代码如下所示。

```
{
  "devDependencies": {
    "babel-cli": "^6.0.0"
  }
}
```

运行 Shell，将 babel-cli 安装在项目中，代码如下所示。

```
npm install --save-dev babel-cli
```

全局安装使用下面的命令：

```
npm install --global babel-cli
```

在 package.json 里添加 npm 的脚本，代码如下所示。

```
{
  "name": "my-project",
  "version": "1.0.0",
  "scripts": {
    "build": "babel src -d lib",
    "bt": "babel es6fortest.js --out-file compiledEs6fortest.js",
    "bast": "babel-node usebabelcore.js"
  },
```

```
"devDependencies": {
  "babel-cli": "^6.26.0",
  "babel-preset-env": "^1.6.1",
  "babel-preset-es2015": "^6.24.1"
}
}
```

上面代码中的 scripts 是快捷命令的映射表。映射表的值记录了真实执行的命令，比如快捷命令 build 对应的映射表的值，即真实执行的命令是 babel src -d lib。

在 Shell 中执行如下命令：

```
npm run build
```

直接指定命令的路径：

```
./node_modules/.bin/babel src -d lib
```

我们还创建了 bast 快捷操作，对应的是 babel-node usebabelcore.js。usebabelcore 文件里的内容如下：

```
let core = require('babel-core').transformFileSync('abs.js', {"presets":
["env"]});
  let ast = core.ast.program.body;
  let code = core.code;
  let prettyAst = JSON.stringify(ast, null, 4);
  console.log(prettyAst);
```

以上涉及 babel-core 的核心接口库的使用，transformFileSync 能够读取一个 JavaScript 文件并对文件内代码进行转译，转译的设置包含 ES 规则的指定等。prettyAst 的内容和 astexplorer 输出的 JSON 是一样的。执行以下代码：

```
npm run bast
```

接下来，创建一个 Swift 工程，写出调用 Shell 的函数。

```
func shell(_ args: String...) -> String {
    let process = Process()
    process.launchPath = "/usr/bin/env"
    process.arguments = args

    let pipe = Pipe()
    process.standardOutput = pipe

    process.launch()
    process.waitUntilExit()

    let data = pipe.fileHandleForReading.readDataToEndOfFile()
    let output: String = String(data: data, encoding: .utf8)!

    return output
}
```

传入先前执行的 Shell 命令作为参数，通过 Shell 函数获得 AST JSON 字符串的返回值。

```
let json = shell("npm","run","bast")
print("ls result:\n\(result)")
```

把 JSON 转成 Dictionary，以供 Swift 使用。或者使用 SwiftyJSON，通过内部自动处理来简化 Dictionary 烦琐的 optional 处理。

```
let jsonStringClear = json.replacingOccurrences(of: "\n", with: "")
let jsonData = jsonStringClear.data(using: .utf8)!

do {
    let dic = try JSONSerialization.jsonObject(with: jsonData, options:
JSONSerialization.ReadingOptions(rawValue: 0)) as! [Dictionary<String, Any>]
    // 对字典进行操作
    ...
} catch let error as NSError { print(error) }
```

1.11 用 Swift 开发一个简单的解释器

1.11.1 四则运算

编写解释器很容易，但如果一开始就抱着实现一门复杂语言的全部高级特性的想法，那么就会陷入目标过大的误区，或者遇到语言设计本身的问题，以致无法完成基本的工作。笔者以前做无用方法删除时，开始就不是抱着全部解析的想法，而是只对需要的部分进行解析。因此，在短时间内就可以看到成果，其他的在之后再一点点完善，这样能够快速获得成就感，就不会轻易放弃了。

因此，要实现一个解释器，就要从最基本的需求切入，到后面再一点一点地加入更多的语言特性。只要保证每一步执行正确，写出一个功能完善的解释器就不是难事了。

我们的第一个目标就是实现一个加法表达式。在上一节中讲到对于一个代码的解析需要先进行分词。本节我们对简单的代码分词（比如 8+3，包含了数字和运算符加号）的 token 结构进行设计。其他情况先都当成文件内容，设置为 eof 类型。

```
public enum OCToken {
    case constant(OCConstant)
    case operation(OCOperation)
    case eof
}

public enum OCConstant {
    case integer(Int)
    case float(Float)
    case boolean(Bool)
    case string(String)
}

public enum OCOperation {
    case plus
}
```

Swift 的枚举可以储存其他类型的关联值和成员值，并且可以在使用时修改这个关联值。这个关联值的类型可以是任意的类型，每个枚举成员关联值的类型也可以不同。其他语言中也有类似的概念，比如 F# 里的基本类型 Discriminated Union，可以查看官方文档了解具体内容。在该类型中，每个值既可以是单个值，也可以是聚合了相同或不同类型的多字段元组，这一点和 Swift 枚举关联值的概念是一致的。

在上面的代码中，定义了一个 OCToken 的枚举，枚举里的 constant 是枚举关联值，constant 的关联值 OCConstant 同时也是枚举关联值。既是关联值，又是枚举关联值的设计特别适合对需要记录的值有多层级分类要求的情况。在对括号进行处理时，关联值可以重复使用，并且减少 OCToken 成员数量。operation 和 constant 一样，operation 的关联值也是一个枚举关联值。设置 operation 和 constant 值的方法如下所示。

```
OCToken.constant(.float(8))
OCToken.operation(.plus)
```

在 switch 的 case 分支里可以使用 let 或 var 提取关联值。

```
case let .constant(.integer(result)):
   eat(.constant(.integer(result)))
```

按照一个字符接一个字符的顺序来解析。

```
func advance() {
   currentIndex += 1
   guard currentIndex < text.count else {
      return
   }
   currentCharacter = text[text.index(text.startIndex, offsetBy: currentIndex)]
}
```

设置属性用来记录当前下标、当前字符，以及当前的 OCToken。

```
private let text: String
private var currentIndex: Int
private var currentCharacter: Character

private var currentTk: OCToken

public init(_ input: String) {
   if input.count == 0 {
      fatalError("Error! input can't be empty")
   }
   self.text = input
   currentIndex = 0
   currentCharacter = text[text.startIndex]
   currentTk = .eof
}
```

通过 nextTk 函数可以获取 OCToken，得到当前下标，通过下标获得当前的字符。判断得到的类型是数字还是加号，不同的类型返回不同的 OCToken。

```
func nextTk() -> OCToken {
```

```
    if currentIndex > self.text.count - 1 {
        return .eof
    }
    
    if CharacterSet.decimalDigits
      .contains(currentCharacter.unicodeScalars.first!) {
        let tk = OCToken.constant(.integer(Int(String(
          currentCharacter))!))
        advance()
        return tk
    }
    
    if currentCharacter == "+" {
        advance()
        return .operation(.plus)
    }
    advance()
    return .eof
}
```

假设在代码中写下 "8+3"，第一个字符是 8，从 CharacterSet 的 decimalDigits 集合就可以判断其是否是数字类型。这里先不处理多位和小数点的情况。如果符合数字类型，那么就返回该类型的 token。通过前面定义好的 advance 将当前字符 currentCharacter 指向后面一个字符。如果第二个字符是加号，则只要判断其是否和加号一样即可。记得调用 advance 函数。第三个字符是 3，也是数字，进入和第一个字符一样的条件语句里处理。

接下来，expr 函数会将加法运算分为左右两部分进行解释运算，并按照一个 token 接一个 token 的顺序来处理，碰到第一个满足整数类型的 token 时就是 left 枚举值，运算符加号的 token 就直接将下一个 token 设置为当前 token。再碰到整数 token 时就是 right 枚举值，最后将 left 枚举值和 right 枚举值进行相加。

```
public func expr() -> Int {
    currentTk = nextTk()
    
    guard case let .constant(.integer(left)) = currentTk else {
        return 0
    }
    eat(currentTk)
    
    eat(.operation(.plus))
    
    guard case let .constant(.integer(right)) = currentTk else {
        return 0
    }
    eat(currentTk)
    
    return left + right
}
```

可以看出，在解析过程中对每个 token 的处理都是预先定好的，并且能够同时检查所解析的代码是否满足语法要求。这归功于 eat 函数，如下所示。

```
private func eat(_ token: OCToken) {
    if currentTk == token {
        currentTk = nextTk()
    } else {
        fatalError("Error: eat wrong")
    }
}
```

eat 函数会将参数里期望的 OCToken 类型和当前的 OCToken 进行对比，如果相同，就将当前的 OCToken 设置为下一个 OCToken。比较两个枚举是否相等时，是没法直接使用 == 运算符的，如果想使用等号比较是否相等，则需要扩展枚举支持 Equatable 协议，同时重载 == 运算符。要实现 eat 函数里对 OCToken 是否相同的判断，就要扩展前面定义的 OCToken、OCConstant 和 OCOperation 三个枚举类型，代码如下所示。

```
extension OCConstant: Equatable {
    public static func == (lhs: OCConstant, rhs: OCConstant) -> Bool {
        switch (lhs, rhs) {
        case let (.integer(left), .integer(right)):
            return left == right
        case let (.float(left), .float(right)):
            return left == right
        case let (.boolean(left), .boolean(right)):
            return left == right
        case let (.string(left), .string(right)):
            return left == right
        default:
            return false
        }
    }
}

extension OCOperation: Equatable {
    public static func == (lhs: OCOperation, rhs: OCOperation) -> Bool {
        switch (lhs, rhs) {
        case (.plus, .plus):
            return true
        default:
            return false
        }
    }
}

extension OCToken: Equatable {
    public static func == (lhs: OCToken, rhs: OCToken) -> Bool {
        switch (lhs, rhs) {
        case let (.constant(left), .constant(right)):
            return left == right
        case let (.operation(left), .operation(right)):
            return left == right
        case (.eof, .eof):
            return true
        default:
            return false
```

```
        }
      }
    }
```

  Equatable 协议是 Swift 的基础协议,Comparable 和 Hashable 都继承于 Equatable 协议。要遵循 Equatable 协议,就必须实现 == 运算符的函数,所以不管是枚举还是结构体,在遵循这个协议后都可以定制自己的相等比较函数,和 Swift 内置的遵循了 Equatable 协议的类型一样能够直接进行相等比较,内置类型包括 Int、Double 和 Float 等。对于类实例和元类型,则可以通过 ObjectIdentifier 函数得到它们的唯一标识,以此判断值是否相等。注意,结构体、枚举、函数或元组是没法使用 ObjectIdentifier 函数的。

  随着 token 类型不断地增加或被修改,实现 Equatable 协议将会是一项枯燥、烦琐且容易出错的重复性工作。解决办法是使用元编程(Meta-programming),目前已有 Swift 代码生成工具 Sourcery 能够解决这个问题。Sourcery 是建立在苹果公司自己的 SourceKit 之上的,使用专为 Swift 设计的模板语言 Stencil 来编写 Sourcery 模板。如果不想使用 Stencil,而是直接使用 SourceKit 来编写工具,那么可以使用 SourceKitten。Stencil 可以使用 StencilSwiftKit 为 Stencil 提供更多节点和 Filter 的库。通过扩展语言的抽象性自动生成样板代码,摆脱重复工作。自动生成样板代码还可以应用到更多的领域,比如 JSON 编码、NSCoding 的实现、Codable 协议的实现、Hashable 协议的实现 和 struct 的初始化,等等。

  接下来,我们来看一看如何调用解释器检验结果。

```
let interperter = OCInterpreter("3+8")
let result = interperter.expr()
print(result)
```

  传入 3+8 表达式,就会输出 11。至此,一个解释器的流程就完成了。但是对于一个完美的解释器来说,还缺少太多的东西。下面我们就来加上这些功能。首先加上对空白字符的支持,实现处理输入字符串里的空格和换行的功能。添加以下函数处理连续的空白字符。

```
private func skipWhiteSpaceAndNewLines() {
    while CharacterSet.whitespacesAndNewlines
      .contains(currentCharacter.unicodeScalars.first!) {
        advance()
    }
}
```

  Swift 里的 CharacterSet 可以表示一组 Unicode 的集合,其提供了 whitespacesAndNewlines 方便我们判断空格和换行。

  有了 skipWhiteSpaceAndNewLines 函数,我们就可以在 nextTk 函数里加上对空格、换行的判断和对 skipWhiteSpaceAndNewLines 函数的调用,如下所示。

```
if CharacterSet.whitespacesAndNewlines
  .contains(currentCharacter.unicodeScalars.first!) {
    skipWhiteSpaceAndNewLines()
    return .whiteSpaceAndNewLine
}
```

  返回的 OCToken 是 whitespaceAndNewLine 类型。不要忘记在 OCToken 的扩展里加上这

个类型的 Equatable 协议的实现。最后在 eat 函数里加上 whitespaceAndNewLine 类型的 token 处理，至此，对于空格、换行的处理就完成了。

现在已完成的功能还只支持个位整数，让我们加上对多位整数的支持吧。多位整数可以采用类似处理空格时的方法，即连续遇到满足是数字这个条件的字符时对其进行组合。首先添加一个对数字处理的函数，如下所示。

```
// 数字处理
private func number() -> OCToken {
    var numStr = ""
    while let character = currentCharacter, CharacterSet.decimalDigits
      .contains((character.unicodeScalars.first!)) {
        numStr += String(character)
        advance()
    }
    
    return .constant(.integer(Int(numStr)!))
}
```

输入 31 + 8 多位整数的表达式，输出结果 39。numStr 是用来累积数字的字符串，返回时通过 Int 强制转换为整型。通过 while 的条件配合 advance 函数，不断获取下一个满足条件的数字字符，累积成多位整数。

浮点数该怎么处理呢？浮点数和整数在字符层面上只是小数点的区别。所以，在处理完数字跳出 while 后，再看看是不是碰到了小数点，如果是，就继续添加小数点后面的数字，具体实现代码如下：

```
// 数字处理
private func number() -> OCToken {
    var numStr = ""
    while let character = currentCharacter, CharacterSet.decimalDigits
      .contains(character.unicodeScalars.first!) {
        numStr += String(character)
        advance()
    }
    
    if let character = currentCharacter, character == "." {
        numStr += "."
        advance()
        while let character = currentCharacter, CharacterSet.decimalDigits
          .contains(character.unicodeScalars.first!) {
            numStr += String(character)
            advance()
        }
        return .constant(.float(Float(numStr)!))
    }
    
    return .constant(.integer(Int(numStr)!))
}
```

四则运算除了加法，还有减法、乘法和除法。

先加上对减法的支持。在 OCOperation 里加上 minus 类型，然后在 Equatable 和 nextTk

里加上对 minus 类型的处理。最后更新一下 expr 函数，并完善 left 枚举值和 right 枚举值的不同运算。目前 expr 函数除了用于做词法分析，还用于做解释的工作。四则运算的代码如下：

```
import Foundation

public enum OCConstant {
    case integer(Int)
    case float(Float)
    case boolean(Bool)
    case string(String)
}

public enum OCOperation {
    case plus
    case minus
    case mult
    case intDiv
}

public enum OCToken {
    case constant(OCConstant)
    case operation(OCOperation)
    case eof
    case whiteSpaceAndNewLine
}

extension OCConstant: Equatable {
    public static func == (lhs: OCConstant, rhs: OCConstant) -> Bool {
        switch (lhs, rhs) {
        case let (.integer(left), .integer(right)):
            return left == right
        case let (.float(left), .float(right)):
            return left == right
        case let (.boolean(left), .boolean(right)):
            return left == right
        case let (.string(left), .string(right)):
            return left == right
        default:
            return false
        }
    }
}

extension OCOperation: Equatable {
    public static func == (lhs: OCOperation, rhs: OCOperation) -> Bool {
        switch (lhs, rhs) {
        case (.plus, .plus):
            return true
        case (.minus, .minus):
            return true
        case (.mult, .mult):
            return true
```

```swift
        case (.intDiv, .intDiv):
            return true
        default:
            return false
        }
    }
}

extension OCToken: Equatable {
    public static func == (lhs: OCToken, rhs: OCToken) -> Bool {
        switch (lhs, rhs) {
        case let (.constant(left), .constant(right)):
            return left == right
        case let (.operation(left), .operation(right)):
            return left == right
        case (.eof, .eof):
            return true
        case (.whiteSpaceAndNewLine, .whiteSpaceAndNewLine):
            return true
        default:
            return false
        }
    }
}

public class OCInterpreter {
    private let text: String
    private var currentIndex: Int
    private var currentCharacter: Character?
    
    private var currentTk: OCToken
    
    public init(_ input: String) {
        if input.count == 0 {
            fatalError("Error! input can't be empty")
        }
        self.text = input
        currentIndex = 0
        currentCharacter = text[text.startIndex]
        currentTk = .eof
    }
    
    // 流程函数
    func nextTk() -> OCToken {
        if currentIndex > self.text.count - 1 {
            return .eof
        }
        
        if CharacterSet.whitespacesAndNewlines
            .contains((currentCharacter?.unicodeScalars.first!)!) {
            skipWhiteSpaceAndNewLines()
            return .whiteSpaceAndNewLine
        }
```

```swift
        if CharacterSet.decimalDigits
          .contains((currentCharacter?.unicodeScalars.first!)!) {
            return number()
        }
        
        if currentCharacter == "+" {
            advance()
            return .operation(.plus)
        }
        if currentCharacter == "-" {
            advance()
            return .operation(.minus)
        }
        if currentCharacter == "*" {
            advance()
            return .operation(.mult)
        }
        if currentCharacter == "/" {
            advance()
            return .operation(.intDiv)
        }
        advance()
        return .eof
    }
    
    public func expr() -> Int {
        currentTk = nextTk()
        
        guard case let .constant(.integer(left)) = currentTk else {
            return 0
        }
        eat(currentTk)
        
        let op = currentTk
        eat(currentTk)
        
        guard case let .constant(.integer(right)) = currentTk else {
            return 0
        }
        eat(currentTk)
        
        if op == .operation(.plus) {
            return left + right
        } else if op == .operation(.minus) {
            return left - right
        } else if op == .operation(.mult) {
            return left * right
        } else if op == .operation(.intDiv) {
            return left / right
        }
        return left + right
    }
}

// 数字处理
```

```swift
    private func number() -> OCToken {
        var numStr = ""
        while let character = currentCharacter, CharacterSet.decimalDigits
          .contains((character.unicodeScalars.first!)) {
            numStr += String(character)
            advance()
        }

        return .constant(.integer(Int(numStr)!))
    }

    // 辅助函数
    private func advance() {
        currentIndex += 1
        guard currentIndex < text.count else {
            currentCharacter = nil
            return
        }
        currentCharacter = text[text.index(text.startIndex, offsetBy: currentIndex)]
    }

    // 在 currentIndex 的值不变的情况下，获取前一个字符
    private func peek() -> Character? {
        let peekIndex = currentIndex + 1
        guard peekIndex < text.count else {
            return nil
        }
        return text[text.index(text.startIndex, offsetBy: peekIndex)]
    }

    private func skipWhiteSpaceAndNewLines() {
        while let character = currentCharacter,
          CharacterSet.whitespacesAndNewlines
           .contains((character.unicodeScalars.first!)) {
            advance()
        }
    }

    private func eat(_ token: OCToken) {
        if currentTk == token {
            currentTk = nextTk()
            if currentTk == OCToken.whiteSpaceAndNewLine {
                currentTk = nextTk()
            }
        } else {
            error()
        }
    }

    func error() {
        fatalError("Error!")
    }
}
```

1.11.2 算术表达式

1.11.1 节中的四则运算只支持两个整数的运算。本节将会讲解支持更多数量的整数运算，比如 31 + 2 - 8 + 10，乘法和除法因为涉及优先级，所以之后再处理。我们先看一个有关加法、减法的语法规则。

下图里的 term 表示运算表达式中的整数。如果加、减整数后出现的又是加、减运算，那么就会循环这个过程，直到不满足加、减条件为止。依据规则，如果只有加、减符号，肯定是不满足的，开头是数字，后面只有一个加号或者减号也不行。

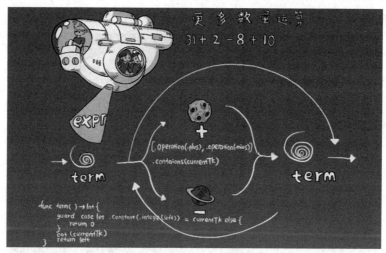

加、减语法规则的实现必然是在语法解析里完成的。语法解析是通过 expr 函数来实现的，下面我们来重写一下 expr 函数，如下所示。

```
public func expr() -> Int {
    currentTk = nextTk()
    
    var result = term()
    
    while [.operation(.plus), .operation(.minus)].contains(currentTk) {
        let tk = currentTk
        eat(currentTk)
        if tk == .operation(.plus) {
            result = result + self.term()
        } else if tk == .operation(.minus) {
            result = result - self.term()
        }
    }
    
    return result
}

// 语法解析中对数字的处理
private func term() -> Int {
    guard case let .constant(.integer(left)) = currentTk else {
        return 0
```

```
        }
        eat(currentTk)
        return left
}
```

我们将对整数的解析封装成以 term 命名的函数，expr 函数遵照语法规则，整数后面只能跟加号或减号，while 条件对整数后的 token 类型做了限制。

在讲解乘法和除法之前，我们先介绍一下优先级。算术运算在优先级相同时是按照左结合运算的。如果遇到下一个运算符的优先级更高的情况，第一个或已运算完的数字就会先和后面的数字进行运算，直至运算符优先级相同，再返回进行左结合运算。加上乘法和除法，以及优先级低的四则运算对应语法，如下图所示。

我们先实现上图中的 factor 函数和 term 函数，其实就是把上一示例中 term 函数里的代码放到 factor 函数中，再把乘法、除法加到 term 函数代码的 while 循环语句中，直到遇到非乘、除运算后停止 while 循环。

```
private func term() -> Int {
    var result = factor()

    while [.operation(.mult), .operation(.intDiv)].contains(currentTk) {
        let tk = currentTk
        eat(currentTk)
        if tk == .operation(.mult) {
            result = result * factor()
        } else if tk == .operation(.intDiv) {
            result = result / factor()
        }
    }
    return result
}

private func factor() -> Int {
    guard case let .constant(.integer(result)) = currentTk else {
        return 0
```

```
        }
        eat(currentTk)
        return result
}
```

当输入 31 + 8 / 2 表达式后，解释器碰到 8 后，会先用 8 除以 2，最后和 31 相加得到 35。目前的算法还没法处理有小括号嵌套的情况，比如 31 + (4 + 5 - (3 + 3)) * 4 - (1 + (51 - 4))，我们先增加有左括号和右括号这两个类型的 token。修改一下 factor 函数的语法规则，使其能够支持左右括号和嵌套，如下图所示。修改后，expr 函数的规则规定 factor 函数碰到左括号还会再递归使用 expr 函数的规则。

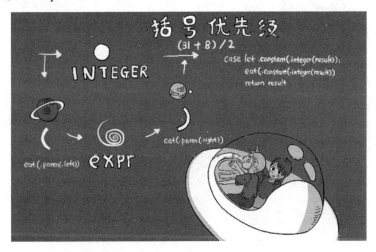

以 (31 + 8) / 2 为例，分析如下图所示。

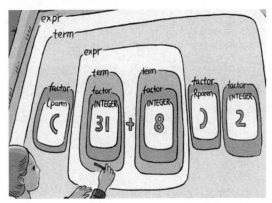

将 factor 函数修改成既可以解析整数，又可以解析括号的表达式，代码如下所示。

```
private func factor() -> Int {
    let tk = currentTk
    switch tk {
    case let .constant(.integer(result)):
        eat(.constant(.integer(result)))
        return result
    case .paren(.left):
        eat(.paren(.left))
```

```
            let result = expr()
            eat(.paren(.right))
            return result
        default:
            return 0
    }
}
```

至此，任意深度嵌套的括号表达式就都能够支持了。下面是完整的代码：

```
import Foundation

public enum OCConstant {
    case integer(Int)
    case float(Float)
    case boolean(Bool)
    case string(String)
}

public enum OCOperation {
    case plus
    case minus
    case mult
    case intDiv
}

public enum OCDirection {
    case left
    case right
}

public enum OCToken {
    case constant(OCConstant)
    case operation(OCOperation)
    case paren(OCDirection)
    case eof
    case whiteSpaceAndNewLine
}

extension OCConstant: Equatable {
    public static func == (lhs: OCConstant, rhs: OCConstant) -> Bool {
        switch (lhs, rhs) {
        case let (.integer(left), .integer(right)):
            return left == right
        case let (.float(left), .float(right)):
            return left == right
        case let (.boolean(left), .boolean(right)):
            return left == right
        case let (.string(left), .string(right)):
            return left == right
        default:
            return false
        }
    }
}
```

```swift
    }

extension OCOperation: Equatable {
    public static func == (lhs: OCOperation, rhs: OCOperation) -> Bool {
        switch (lhs, rhs) {
        case (.plus, .plus):
            return true
        case (.minus, .minus):
            return true
        case (.mult, .mult):
            return true
        case (.intDiv, .intDiv):
            return true
        default:
            return false
        }
    }
}

extension OCDirection: Equatable {
    public static func == (lhs: OCDirection, rhs: OCDirection) -> Bool {
        switch (lhs, rhs) {
        case (.left, .left):
            return true
        case (.right, .right):
            return true
        default:
            return false
        }
    }
}

extension OCToken: Equatable {
    public static func == (lhs: OCToken, rhs: OCToken) -> Bool {
        switch (lhs, rhs) {
        case let (.constant(left), .constant(right)):
            return left == right
        case let (.operation(left), .operation(right)):
            return left == right
        case (.eof, .eof):
            return true
        case (.whiteSpaceAndNewLine, .whiteSpaceAndNewLine):
            return true
        case let (.paren(left), .paren(right)):
            return left == right
        default:
            return false
        }
    }
}

public class OCLexer {
    private let text: String
    private var currentIndex: Int
```

```swift
    private var currentCharacter: Character?

    public init(_ input: String) {
        if input.count == 0 {
            fatalError("Error! input can't be empty")
        }
        self.text = input
        currentIndex = 0
        currentCharacter = text[text.startIndex]
    }

    // 流程函数
    func nextTk() -> OCToken {
        if currentIndex > self.text.count - 1 {
            return .eof
        }

        if CharacterSet.whitespacesAndNewlines
          .contains((currentCharacter?.unicodeScalars.first!)!) {
            skipWhiteSpaceAndNewLines()
            return .whiteSpaceAndNewLine
        }

        if CharacterSet.decimalDigits
          .contains((currentCharacter?.unicodeScalars.first!)!) {
            return number()
        }

        if currentCharacter == "+" {
            advance()
            return .operation(.plus)
        }
        if currentCharacter == "-" {
            advance()
            return .operation(.minus)
        }
        if currentCharacter == "*" {
            advance()
            return .operation(.mult)
        }
        if currentCharacter == "/" {
            advance()
            return .operation(.intDiv)
        }
        if currentCharacter == "(" {
            advance()
            return .paren(.left)
        }
        if currentCharacter == ")" {
            advance()
            return .paren(.right)
        }
        advance()
        return .eof
```

```swift
        }
        // 数字处理
        private func number() -> OCToken {
            var numStr = ""
            while let character = currentCharacter, CharacterSet.decimalDigits
                .contains((character.unicodeScalars.first!)) {
                numStr += String(character)
                advance()
            }
            
            return .constant(.integer(Int(numStr)!))
        }
        
        // 辅助函数
        private func advance() {
            currentIndex += 1
            guard currentIndex < text.count else {
                currentCharacter = nil
                return
            }
            currentCharacter = text[text.index(text.startIndex, offsetBy: currentIndex)]
        }
        
        //在 currentIndex 的值不变的情况下，获取前一个字符
        private func peek() -> Character? {
            let peekIndex = currentIndex + 1
            guard peekIndex < text.count else {
                return nil
            }
            return text[text.index(text.startIndex, offsetBy: peekIndex)]
        }
        
        private func skipWhiteSpaceAndNewLines() {
            while let character = currentCharacter, CharacterSet.whitespacesAndNewlines.contains((character.unicodeScalars.first!)) {
                advance()
            }
        }
    }
    
    public class OCInterpreter {
        
        private var lexer: OCLexer
        private var currentTk: OCToken
        
        public init(_ input: String) {
            lexer = OCLexer(input)
            currentTk = lexer.nextTk()
        }
        
        public func expr() -> Int {
            
            var result = term()
```

```swift
        while [.operation(.plus), .operation(.minus)]
          .contains(currentTk) {
            let tk = currentTk
            eat(currentTk)
            if tk == .operation(.plus) {
                result = result + self.term()
            } else if tk == .operation(.minus) {
                result = result - self.term()
            }
        }
        
        return result
    }
    
    // 语法解析中对数字的处理
    private func term() -> Int {
        var result = factor()
        
        while [.operation(.mult), .operation(.intDiv)]
          .contains(currentTk) {
            let tk = currentTk
            eat(currentTk)
            if tk == .operation(.mult) {
                result = result * factor()
            } else if tk == .operation(.intDiv) {
                result = result / factor()
            }
        }
        return result
    }
    
    private func factor() -> Int {
        let tk = currentTk
        switch tk {
        case let .constant(.integer(result)):
            eat(.constant(.integer(result)))
            return result
        case .paren(.left):
            eat(.paren(.left))
            let result = expr()
            eat(.paren(.right))
            return result
        default:
            return 0
        }
    }
    
    private func eat(_ token: OCToken) {
        if currentTk == token {
            currentTk = lexer.nextTk()
            if currentTk == OCToken.whiteSpaceAndNewLine {
                currentTk = lexer.nextTk()
            }
```

```
        } else {
            error()
        }
    }

    func error() {
        fatalError("Error!")
    }
}
```

至此,一个完整的算术表达式解释器就完成了。大家可以试着输入 31 + (4 + 5 - (3 + 3)) * 4 - (1 + (51 - 4)),看一看这个解释器的结果和自己手动计算的结果是不是一样的。

1.11.3 中间表示

在 1.11.2 节中,我们让解释器能够处理四则运算表达式,但是对于 Objective-C 这样完整的语言结构来说,我们还需要使用一个更加复杂的中间表示结构来支撑。中间表示(Intermediate Representation)也可称为"IR"。程序代码在词法、语法阶段构建 IR。解释器依据 IR 输入来进行解释执行。

我们使用树形结构来构建 IR。比如 31 + (8 / 2),使用树形结构进行中间表示的方式有两种:一种是根据具体语言的语法规则来表示的分析树;另一种是抽象语法树,简称"AST"。分析树和 AST 的区别如下图所示。

可以看出,AST 更适合作为 IR 的结构,优先级越高的运算符会在 AST 越低的位置。AST 上的各个元素称为"节点"。我们现在用程序来表现 AST,先创建一个 AST 的协议,让所有的节点都遵循这个协议。

```
public protocol OCAST {}
```

四则运算的运算符可以合并成一个运算符节点类 BinOp,该类能够支持左右整数的运算。

```
class OCBinOp: OCAST {
    let left: OCAST
    let operation: OCBinOpType
```

```
    let right: OCAST

    init(left: OCAST, operation: OCBinOpType, right: OCAST) {
        self.left = left
        self.operation = operation
        self.right = right
    }
}
```

关于运算符节点的设计，王巍（@OneVcat）也做过一种设计，当时微博上有一道面试题考查怎么实现四则运算，王巍优雅地实现了运算节点的设计，实现如下图所示：

对于整型，也可以定义一个类似 OCConstant 那样能存值的枚举结构，代码如下所示。

```
public enum OCNumber: OCAST {
    case integer(Int)
    case float(Float)
}
```

按照以上枚举结构，改造 term 函数、factor 函数和 expr 函数，代码如下所示。

```
public func expr() -> OCAST {
    var node = term()

    while [.operation(.plus), .operation(.minus)].contains(currentTk) {
        let tk = currentTk
        eat(currentTk)
        if tk == .operation(.plus) {
            node = OCBinOp(left: node, operation: .plus, right: term())
        } else if tk == .operation(.minus) {
            node = OCBinOp(left: node, operation: .minus, right: term())
        }
    }
    return node
}
```

```swift
// 语法解析中对数字的处理
private func term() -> OCAST {
    var node = factor()
    
    while [.operation(.mult), .operation(.intDiv)].contains(currentTk) {
        let tk = currentTk
        eat(currentTk)
        if tk == .operation(.mult) {
            node = OCBinOp(left: node, operation: .mult, right: factor())
        } else if tk == .operation(.intDiv) {
            node = OCBinOp(left: node, operation: .intDiv, right: factor())
        }
    }
    return node
}

private func factor() -> OCAST {
    let tk = currentTk
    switch tk {
    case let .constant(.integer(result)):
        eat(.constant(.integer(result)))
        return OCNumber.integer(result)
    case .paren(.left):
        eat(.paren(.left))
        let result = expr()
        eat(.paren(.right))
        return result
    default:
        return OCNumber.integer(0)
    }
}
```

输入 4 + 3 - 2，生成的结构如下所示。

```
▼ 🅛 result = (HTN.OCBinOp) 0x000060c0000c2a00
   ▼ left = (HTN.OCBinOp) 0x000060c0000c2990
      ▼ left = (HTN.OCNumber) integer
         integer = (Int) 4
      operation = (HTN.OCBinOpType) plus
      ▼ right = (HTN.OCNumber) integer
         integer = (Int) 3
   operation = (HTN.OCBinOpType) minus
   ▼ right = (HTN.OCNumber) integer
      integer = (Int) 2
```

可以看出，和我们设计的一样，优先级更高的 4 + 3 在树形结构里离根节点越远。输入 4 + 3 * 2，生成的结构如下所示。

```
▼ 🅛 result = (HTN.OCBinOp) 0x00006000000c1f10
    ▼ left = (HTN.OCNumber) integer
          integer = (Int) 4
       operation = (HTN.OCBinOpType) plus
    ▼ right = (HTN.OCBinOp) 0x00006000000c1ea0
       ▼ left = (HTN.OCNumber) integer
             integer = (Int) 3
          operation = (HTN.OCBinOpType) mult
       ▼ right = (HTN.OCNumber) integer
             integer = (Int) 2
```

此处 3 * 2 的优先级更高，所以离树的根节点更远。

在生成了抽象语法树之后，需要能够对树进行解释并计算出表达式的值。根据这个树的特性——优先级越高，离根节点越远，在计算数字时就应该先计算远的。首先想到的就是后序遍历法，从根节点开始递归，从左到右访问子节点，到达最远节点开始计算。

按照后序遍历访问并解释运算的函数，代码如下所示。

```
// eval
public func eval(node: OCAST) -> OCValue {
    switch node {
    case let number as OCNumber:
        return eval(number: number)
    case let binOp as OCBinOp:
        return eval(binOp: binOp)
    default:
        return .none
    }
}

func eval(number: OCNumber) -> OCValue {
    return .number(number)
}

func eval(binOp: OCBinOp) -> OCValue {
    guard case let .number(leftResult) = eval(node: binOp.left),
        case let .number(rightResult) = eval(node: binOp.right) else {
        fatalError("Error! binOp is wrong")
    }
    
    switch binOp.operation {
    case .plus:
        return .number(leftResult + rightResult)
    case .minus:
        return .number(leftResult - rightResult)
    case .mult:
        return .number(leftResult * rightResult)
    case .intDiv:
        return .number(leftResult / rightResult)
    }
}
```

}

eval 传入前面生成的抽象语法树的结构后，就能够得出四则运算表达式的结果。eval(binOp: OCBinOp) 函数里的 +、-、*、/ 运算符目前仍不支持自定义的 OCNumber 结构运算，所以需要重载运算符。代码如下所示。

```
extension OCNumber {
    // binOp
    static func + (left: OCNumber, right: OCNumber) -> OCNumber {
        switch (left, right) {
        case let (.integer(left), .integer(right)):
            return .integer(left + right)
        case let (.float(left), .float(right)):
            return .float(left + right)
        case let (.float(left), .integer(right)):
            return .float(left + Float(right))
        case let (.integer(left), .float(right)):
            return .float(Float(left) + right)
        }
    }

    static func - (left: OCNumber, right: OCNumber) -> OCNumber {
        switch (left, right) {
        case let (.integer(left), .integer(right)):
            return .integer(left - right)
        case let (.float(left), .float(right)):
            return .float(left - right)
        case let (.float(left), .integer(right)):
            return .float(left - Float(right))
        case let (.integer(left), .float(right)):
            return .float(Float(left) - right)
        }
    }

    static func * (left: OCNumber, right: OCNumber) -> OCNumber {
        switch (left, right) {
        case let (.integer(left), .integer(right)):
            return .integer(left * right)
        case let (.float(left), .float(right)):
            return .float(left * right)
        case let (.float(left), .integer(right)):
            return .float(left * Float(right))
        case let (.integer(left), .float(right)):
            return .float(Float(left) * right)
        }
    }

    static func / (left: OCNumber, right: OCNumber) -> OCNumber {
        switch (left, right) {
        case let (.integer(left), .integer(right)):
            return .integer(left / right)
        case let (.float(left), .float(right)):
```

```
            return .float(left / right)
        case let (.float(left), .integer(right)):
            return .float(left / Float(right))
        case let (.integer(left), .float(right)):
            return .float(Float(left) / right)
        }
    }
}
```

重载后，OCNumber 里的整型和浮点型数字运算的各种组合就都能支持了。输入 4 + 3 * 2，可以得到 number(HTN.OCNumber.integer(10))。

先前我们做的是支持加、减、乘、除这样针对两个操作数的二元运算符，那么对一个操作数进行运算操作的运算符叫什么呢？对，就是"一元运算符"。一元运算符包含一元加运算符和一元减运算符。一元加运算符对操作数没有影响，一元减运算符会对操作数进行求反运算。

接下来，添加一个一元运算符的节点。

```
public enum OCUnaryOperationType {
    case plus
    case minus
}

class OCUnaryOperation: OCAST {
    let operation: OCUnaryOperationType
    let operand: OCAST

    init(operation: OCUnaryOperationType, operand: OCAST) {
        self.operation = operation
        self.operand = operand
    }
}
```

修改 factor 函数来适配更新的规则，在新规则里增加一元运算符。

```
case .operation(.plus):
    eat(.operation(.plus))
    return OCUnaryOperation(operation: .plus, operand: factor())
case .operation(.minus):
    eat(.operation(.minus))
    return OCUnaryOperation(operation: .minus, operand: factor())
```

添加解释执行 OCUnaryOperation 的函数 eval(unaryOperation: OCUnaryOperation)。

```
func eval(unaryOperation: OCUnaryOperation) -> OCValue {
    guard case let .number(result) = eval(node: unaryOperation.operand) else {
        fatalError("Error: eval unaryOperation")
    }
    switch unaryOperation.operation {
    case .plus:
        return .number(+result)
    case .minus:
        return .number(-result)
```

与重载运算符一样,重载一元运算符。

```
// Unary
static prefix func + (left: OCNumber) -> OCNumber {
   switch left {
   case let .integer(value):
      return .integer(+value)
   case let .float(value):
      return .float(+value)
   }
}

static prefix func - (left: OCNumber) -> OCNumber {
   switch left {
   case let .integer(value):
      return .integer(-value)
   case let .float(value):
      return .float(-value)
   }
}
```

同时在 OCAST 非明确节点函数 eval(node: OCAST) 里添加对 OCUnaryOperation 类型节点情况的处理。

```
case let unaryOperation as OCUnaryOperation:
   return eval(unaryOperation: unaryOperation)
```

至此,按照 AST 方式完成了一元运算符的支持。现在输入 4 + - 3 * 2 ,看看结果是否符合预期。

打印的结果是 number(HTN.OCNumber.integer(-2)),和预期一致。

1.11.4 变量

本节将讲解 Objective-C 语法。由于 Objective-C 语法太复杂,所以不会全部讲解。先从最简单的给变量赋值和查找变量值开始说明,类型则会在后面讲解。变量处理起来比其他的语法要复杂,操作过程涉及变量的绑定、取值和作用域(变量可见区域)等。示例代码如下所示。

```
@interface OurClass

@end

@implementation OurClass

- (void)run {
   a = 13;
   ab = (3 + 2) + a * 2;
}
```

```
@end
```

上面的示例中引入了很多 Objective-C 关键字，比如 @interface、@implementation 和 @end，以及很多符号，比如 =、{、}等。

语法规则按照 method 是由多个 statement 组成的，每个 statement 以分号表示结束。变量的 token 类型是 id(string)，值就是变量的名字，例如，上面代码里的 a 和 ab。

我们先来为这些新关键字和符号增加 token 类型。

```
case brace(OCDirection)      // { }
case interface
case end
case implementation
case id(String)              // string
case semi
case assign
```

在 Equatable 协议扩展里添加对应的处理。

```
case let (.brace(left), .brace(right)):
   return left == right
case (.asterisk, .asterisk):
   return true
case (.interface, .interface):
   return true
case (.end, .end):
   return true
case (.implementation, .implementation):
   return true
case let (.id(left), .id(right)):
   return left == right
case (.semi, .semi):
   return true
case (.assign, .assign):
   return true
case (.return, .return):
   return true
```

接下来处理这些新增加的 token 的解析工作。除数字和运算符外，还需要处理后面会进行的语义处理的变量名、方法名和关键字等。增加一个 identifier 类型 的 token。当字符集满足 alphanumerics 时，就开始处理。

```
// identifier
if CharacterSet.alphanumerics
  .contains((currentCharacter.unicodeScalars.first!)!) {
    return id()
}
// id()方法会不断检查后面的字符是否满足 alphanumerics 字符集，同时累加记录字符。在不满足时跳出
// identifier and keywords
private func id() -> OCToken {
    var idStr = ""
    while let character = currentCharacter, CharacterSet.alphanumerics
```

```
        .contains(character.unicodeScalars.first!) {
        idStr += String(character)
        advance()
    }

    // 关键字
    if let token = keywords[idStr] {
        return token
    }

    return .id(idStr)
}
```

Objective-C 里有很多以 @ 开头的关键字,这类关键字可以通过统一的方法处理,字符串的处理和 id() 方法类似,基本上都是根据获取到的关键字返回对应的 token。

```
// @符号的处理
private func at() -> OCToken {
    advance()
    var atStr = ""
    while let character = currentCharacter, CharacterSet.alphanumerics
        .contains((character.unicodeScalars.first!)) {
        atStr += String(character)
        advance()
    }
    if atStr == "interface" {
        return .interface
    }
    if atStr == "end" {
        return .end
    }
    if atStr == "implementation" {
        return .implementation
    }

    fatalError("Error: at string not support")
}
```

新增加的符号在 nextTk() 方法里的处理,如下所示。

```
if currentCharacter == "@" {
    return at()
}
if currentCharacter == ";" {
    advance()
    return .semi
}
if currentCharacter == "=" {
    advance()
    return .assign
}
if currentCharacter == "{" {
    advance()
    return .brace(.left)
```

```
    }
    if currentCharacter == "}" {
        advance()
        return .brace(.right)
    }
    if currentCharacter == "*" {
        advance()
        return .asterisk
    }
```

token 处理完后就可以设计节点了。从根级 OCProgram 开始设计。

```
class OCProgram: OCAST {
    let interface: OCInterface
    let implementation: OCImplementation
    init(interface:OCInterface, implementation: OCImplementation) {
        self.interface = interface
        self.implementation = implementation
    }
}
```

上面的设计中包含了 OCInterface 和 OCImplementation。OCInterface 会在讲作用域的处理时用到。OCImplementation 里有一个包含了 OCMethod 的集合。

```
class OCImplementation: OCAST {
    let name: String
    let methodList: [OCMethod]
    init(name: String, methodList: [OCMethod]) {
        self.name = name
        self.methodList = methodList
    }
}
```

OCMethod 的定义如下：

```
class OCMethod: OCAST {
    let returnIdentifier: String
    let methodName: String
    let statements: [OCAST]
    init(returnIdentifier: String, methodName: String, statements:[OCAST])
    {
        self.returnIdentifier = returnIdentifier
        self.methodName = methodName
        self.statements = statements
    }
}
```

实际上，OCMethod 需要处理的情况比上面代码里处理的情况要多，比如参数节点和返回类型节点。本示例主要是想把关注点放到方法里的语句集合上，其他的语法支持可以在以后完善。

我们再来看看示例中 Objective-C 的方法代码：

```
- (void)run {
    a = 13;
```

```
    ab = (3 + 2) + a * 2;
}
```

以上方法中的两个语句都是赋值语句。根据赋值语句的规则，assign 的左边是变量，assign 的右边是表达式。赋值语句是由左边的变量和右边的表达式组成的。赋值语句的节点设计如下：

```
class OCAssign: OCAST {
    let left: OCVar
    let right: OCAST

    init(left: OCVar, right: OCAST) {
        self.left = left
        self.right = right
    }
}
```

变量的节点 OCVar 只需要记录变量名即可。OCParser 新增加了 interface()、implementation()、methodList()、method()、statements()、statement()、assignStatement()、variable() 方法。这些方法会将 token 集合转换成新的节点，新节点构成新的语法树，具体实现如下所示。

```
private func program() -> OCProgram {
    return OCProgram(interface: interface(), implementation: implementation())
}

private func interface() -> OCInterface {
    eat(.interface)
    guard case let .id(name) = currentTk else {
        fatalError("Error interface")
    }
    eat(.id(name))
    eat(.end)
    return OCInterface(name: name)
}

private func implementation() -> OCImplementation {
    eat(.implementation)
    guard case let .id(name) = currentTk else {
        fatalError("Error implementation")
    }
    eat(.id(name))
    let methodListNode = methodList()
    eat(.end)
    return OCImplementation(name: name, methodList: methodListNode)
}

private func methodList() -> [OCMethod] {
    var methods = [OCMethod]()
    while currentTk == .operation(.plus) || currentTk == .operation(.minus) {
        eat(currentTk)
        methods.append(method())
    }
}
```

```swift
        return methods
    }
    
    private func method() -> OCMethod {
        eat(.paren(.left))
        guard case let .id(reStr) = currentTk else {
            fatalError("Error reStr")
        }
        eat(.id(reStr))
        eat(.paren(.right))
        guard case let .id(methodName) = currentTk else {
            fatalError("Error methodName")
        }
        eat(.id(methodName))
        eat(.brace(.left))
        let statementsNode = statements()
        eat(.brace(.right))
        return OCMethod(returnIdentifier: reStr, methodName: methodName, statements: statementsNode)
    }
    
    private func statements() -> [OCAST] {
        let sNode = statement()
        var statements = [sNode]
        while currentTk == .semi {
            eat(.semi)
            statements.append(statement())
        }
        return statements
    }
    
    private func statement() -> OCAST {
        switch currentTk {
        case .id:
            return assignStatement()
        default:
            return empty()
        }
    }
    
    private func assignStatement() -> OCAssign {
        let left = variable()
        eat(.assign)
        let right = expr()
        return OCAssign(left: left, right: right)
    }
    
    private func variable() -> OCVar {
        guard case let .id(name) = currentTk else {
            fatalError("Error: var was wrong")
        }
        eat(.id(name))
        
        return OCVar(name: name)
```

在 methodList() 方法中的 implementation 里，如果碰到 + 或 -，那么就要开始对 method() 进行解析了。目前 OCMethod 结构还没有记录类方法和实例方法的属性。更完备的解析需求可以根据自己的情况添加，目前我们的主要任务是尽快完成解释器的雏形和框架，方便以后实现对更多语法特性的支持。

注意，要更新一下以前的 factor() 方法。这里只需要在 switch case 的 default 里让其按照 variable() 方法解析节点即可。返回的节点会对应到表达式节点对应的子节点里。

再看一下 OCInterpreter 解释器对于增加变量节点后的支持。新增了两个属性，一个是 ast，记录 Parser 返回的抽象语法树；另一个是 scopes，用来保存变量。这样可以在变量赋值时添加变量值到 scopes 里，并且在访问变量时能够从 scopes 里取出值来。

两个属性定义如下所示。

```
private let ast: OCAST
private var scopes: [String: OCValue]
```

scopes 使用的是一个字典结构，键是变量名，值是对应的变量值。解释器对应的解释代码如下所示。

```
// 赋值
func eval(assign: OCAssign) -> OCValue {
    scopes[assign.left.name] = eval(node: assign.right)
    return .none
}

// 访问变量
func eval(variable: OCVar) -> OCValue {
    guard let value = scopes[variable.name] else {
        fatalError("Error: eval var")
    }
    return value
}
```

为了能够让解释器完整地访问 OCParser 生成的语法树，OCInterpreter 还需要补全对新节点的解释。

```
case let program as OCProgram:
    return eval(program: program)
case let implementation as OCImplementation:
    return eval(implementation: implementation)
case let method as OCMethod:
    return eval(method: method)
case let assign as OCAssign:
    return eval(assign: assign)
case let variable as OCVar:
    return eval(variable: variable)
```

运行后的结果和我们的预期一样，如下所示。

```
scope is:
["a": HTN.OCValue.number(HTN.OCNumber.integer(13)), "ab":
```

```
HTN.OCValue.number(HTN.OCNumber.integer(31))]
```

a 被赋值后，在 ab 赋值的表达式里 a 的值 13 会被带入表达式 (3 + 2) + a * 2，计算结果为 31，完全正确。

注意，这里的 scope 并不是最终的设计。因为只考虑了一个作用域，所以，如果在不同的作用域又使用相同的变量名，那么这个结构就会出问题。比如递归函数，进入递归，扩展一个环境，在返回后外层的变量还需要获取先前的绑定值。现在像 scope 这种以变量名作为键值、变量值为值的字典结构，不管在哪个作用域里，都只记录最新值。现在的设计显然是无法应对这种情况的，应该用什么方法来解决这个问题呢？

方法就是使用堆栈，最内层的绑定会在栈顶，进入内层环境的过程就是压栈过程，这样在查找变量时就可以先找到最内层环境绑定的变量值。

1.11.5 属性

接下来 OCInterface 就要开始发挥作用了，因为这一节就要讲解属性及对注释的处理。Objective-C 代码如下：

```
@interface OurClass
// 定义属性
@property (nonatomic, assign) CGFloat pa;
@property (nonatomic, assign) CGFloat pb;

@end

@implementation OurClass

/* 开始运算 */
- (void)run {
    a = 13.3;
    ab = (3 + 2) + a * 2;
}
```

在了解了@关键字的处理方法之后，对@property 的解析就容易很多。观察上面的代码就会发现，只要在 OCInterface 节点里添加一个属性集合和一个属性节点，就能够完成新节点的设计工作，代码如下：

```
class OCInterface: OCAST {
    let name: String
    let propertyList: [OCPropertyDeclaration]
    init(name: String, propertyList: [OCPropertyDeclaration]) {
        self.name = name
        self.propertyList = propertyList
    }
}

class OCPropertyDeclaration: OCAST {
    let propertyAttributesList: [OCPropertyAttribute]
    let type: String
    let name: String
```

```swift
    init(propertyAttributesList: [OCPropertyAttribute], type: String, name: String) {
        self.propertyAttributesList = propertyAttributesList
        self.type = type
        self.name = name
    }
}

class OCPropertyAttribute: OCAST {
    let name :String
    init(name: String) {
        self.name = name
    }
}
```

在 OCParser 里对新节点的处理，代码如下：

```swift
private func interface() -> OCInterface {
    eat(.interface)
    guard case let .id(name) = currentTk else {
        fatalError("Error interface")
    }
    eat(.id(name))
    let pl = propertyList()
    eat(.end)
    return OCInterface(name: name, propertyList: pl)
}

private func propertyList() -> [OCPropertyDeclaration] {
    var properties = [OCPropertyDeclaration]()
    while currentTk == .property {
        eat(.property)
        eat(.paren(.left))
        let pa = propertyAttributes()
        eat(.paren(.right))
        guard case let .id(pType) = currentTk else {
            fatalError("Error: property type wrong")
        }
        eat(.id(pType))
        guard case let .id(name) = currentTk else {
            fatalError("Error: property name wrong")
        }
        eat(.id(name))
        let pd = OCPropertyDeclaration(propertyAttributesList: pa, type: pType, name: name)
        properties.append(pd)
        eat(.semi)
    }
    return properties
}

private func propertyAttributes() -> [OCPropertyAttribute] {
    let p = propertyAttribute()
    var pa = [p]
```

```
    while currentTk == .comma {
        eat(.comma)
        pa.append(propertyAttribute())
    }
    return pa
}

private func propertyAttribute() -> OCPropertyAttribute {
    guard case let .id(name) = currentTk else {
        fatalError("Error: propertyAttribute wrong")
    }
    eat(.id(name))
    return OCPropertyAttribute(name: name)
}
```

Objective-C 的注释有两种，一种是 // comments，另一种是 /* comments /。第一种是碰到 // 开始注释，第二种是碰到 / 开始注释。我们从 nextTk() 方法的 currentCharacter == "/" 条件入手，编写代码如下：

```
if currentCharacter == "/" {
    // 处理可能的注释的情况
    if peek() == "/" {
        advance()
        advance()
        return commentsFromDoubleSlash()
    } else if peek() == "*" {
        advance()
        advance()
        return commentsFromSlashAsterisk()
    } else {
        advance()
        return .operation(.intDiv)
    }
}
```

当遇到 //注释的结束条件时，要换行。commentsFromDoubleSlash() 方法会对每个字符进行累加记录。当遇到字符集 CharacterSet.newlines 时，就返回完整的注释。代码如下：

```
// 过滤 // 这种注释
private func commentsFromDoubleSlash() -> OCToken {
    var cStr = ""
    while let character = currentCharacter, !CharacterSet.newlines
      .contains(character.unicodeScalars.first!) {
        advance()
        cStr += String(character)
    }
    return .comments(cStr)
}
```

/* 的结束条件是 */，所以在 commentsFromSlashAsterisk() 方法里碰到 * 后，需要再看一下后面一个字符是否是 /。

```
// 过滤 /* */ 这样的注释
```

```
private func commentsFromSlashAsterisk() -> OCToken {
    var cStr = ""
    while let character = currentCharacter {
        if character == "*" && peek() == "/" {
            advance()
            advance()
            break
        } else {
            advance()
            cStr += String(character)
        }
    }
    return .comments(cStr)
}
```

1.11.6 静态检查

本节将创建一个符号表。根据符号表可以检查变量在使用前是否被声明。声明类型后，确保我们能给变量赋正确类型的值。符号分为变量符号和内置类型符号，内置类型符号是可以直接使用的类型符号，比如整型、浮点型、布尔值和字符串等。首先，创建一个符号协议 OCSymbol，让两种符号都遵循这个协议。

```
protocol OCSymbol {
    var name : String { get }
}
```

内置类型符号表使用的是一个枚举类型，协议中的 name 属性根据不同的枚举值返回不同的字符串。

```
public enum OCBuiltInTypeSymbol: OCSymbol {
    case integer
    case float
    case boolean
    case string
    
    var name: String {
        switch self {
        case .integer:
            return "NSUInteger"
        case .float:
            return "CGFloat"
        case .boolean:
            return "BOOL"
        case .string:
            return "NSString"
        }
    }
}
```

在变量符号表中增加一个类型属性，代码如下所示。

```
class OCVariableSymbol: OCSymbol {
    let name: String
```

```
    let type: OCSymbol

    init(name: String, type: OCSymbol) {
        self.name = name
        self.type = type
    }
}
```

再创建一个 OCSymbolTable 符号表的类用来记录变量和类型的对应关系。符号表是专门用来跟踪代码里各种符号的抽象数据类型的。符号表除存储符号外，还有查找符号的功能。

```
public class OCSymbolTable {
    var symbols: [String: OCSymbol] = [:]

    let name: String

    init(name: String) {
        self.name = name
        defineBuiltInTypes()
    }

    private func defineBuiltInTypes() {
        define(OCBuiltInTypeSymbol.integer)
        define(OCBuiltInTypeSymbol.float)
        define(OCBuiltInTypeSymbol.boolean)
        define(OCBuiltInTypeSymbol.string)
    }

    func define(_ symbol: OCSymbol) {
        symbols[symbol.name] = symbol
    }

    func lookup(_ name: String) -> OCSymbol? {
        if let symbol = symbols[name] {
            return symbol
        }
        return nil
    }
}
```

其中，define 方法是用来存储符号的，会将符号的名称作为"键"，符号实例作为"symbols 字典的值"。lookup 方法则是用来查找符号。

我们可以在解析的过程中构建符号表。但是为了将静态分析和解释过程分开、不耦合，我们使用访问者模式创建一个 Visitor 来专门做静态分析。为了方便日后做更多的事情而不用将节点处理逻辑都放在一起，我们采用面向协议编程的设计模式，创建一个 OCVisitor 访问者协议，通过 extension 实现协议中所有方法的默认实现，这些实现主要是保障访问者能够以递归下降方式递归完语法树的各个节点。然后再创建一个 OCStaticAnalyzer 类以遵循 OCVisitor 这个访问者协议，并对那些和构建符号表相关的节点的访问方法进行重写。

OCVisitor 访问者协议的定义如下：

```
protocol OCVisitor: class {
```

```swift
    func visit(node: OCAST)
    func visit(program: OCProgram)
    func visit(interface: OCInterface)
    func visit(propertyDeclaration: OCPropertyDeclaration)
    func visit(propertyAttribute: OCPropertyAttribute)
    func visit(implementation: OCImplementation)
    func visit(method: OCMethod)
    func visit(compoundStatement: OCCompoundStatement)
    func visit(identifier: OCIdentifier)
    func visit(assign: OCAssign)
    func visit(variable: OCVar)
    func visit(number: OCNumber)
    func visit(unaryOperation: OCUnaryOperation)
    func visit(binOp: OCBinOp)
    func visit(noOp: OCNoOp)
}
```

协议中定义了各个节点的访问方法。这些在 extension 中实现的方法只是根据节点类型递归下降以调用对应的访问者方法来访问而已。具体用来构建符号表的实现则是在 OCStaticAnalyzer 里通过重写相关访问方法完成的。

```swift
public class OCStaticAnalyzer: OCVisitor {
    private var symbolTable = OCSymbolTable(name: "global")
    
    public init() {
        
    }
    
    public func analyze(node: OCAST) -> OCSymbolTable {
        visit(node: node)
        return symbolTable
    }
    
    func visit(propertyDeclaration: OCPropertyDeclaration) {
        guard symbolTable.lookup(propertyDeclaration.name) == nil else {
            fatalError("Error: duplicate identifier \(propertyDeclaration.name) found")
        }
        
        guard let symbolType = symbolTable.lookup(propertyDeclaration.type) else {
            fatalError("Error: \(propertyDeclaration.type) type not found")
        }
        
        symbolTable.define(OCVariableSymbol(name: propertyDeclaration.name, type: symbolType))
    }
    
    func visit(variable: OCVar) {
        guard symbolTable.lookup(variable.name) != nil else {
            fatalError("Error: \(variable.name) variable not found")
        }
    }
}
```

```
    func visit(assign: OCAssign) {
        guard symbolTable.lookup(assign.left.name) != nil else {
            fatalError("Error: \(assign.left.name) not found")
        }
    }
}
```

func visit(propertyDeclaration:OCPropertyDeclaration) 的作用是访问属性声明节点,并检查属性名是否重复定义,以及属性类型是否未定义。然后将该属性声明节点的属性添加到符号表中。在访问变量节点时,执行 func visit(variable: OCVar) 方法。func visit(variable:OCVar) 方法会检查变量是否被声明。在给变量赋值时,变量也会做同样的检查。

还记得 1.11.5 节里的示例吗?当时没有在方法里使用属性是因为属性的作用域和方法的作用域不一样,因此,变量就需要支持多级作用域。现在我们就开始写支持多级作用域的代码。先更新一下 Objective-C,代码如下:

```
@interface OurClass
// 定义属性
@property (nonatomic, assign) CGFloat pa;
@property (nonatomic, assign) CGFloat pb;

@end

@implementation OurClass

/* 开始运算 */
- (void)run {
    CGFloat ta = 4.3;
    pb = 13.3;
    pa = (3 + 2) + pb * 2 - ta;
}

@end
```

增加一个变量声明节点 OCVariableDeclaration,用来支持解析 CGFloat ta = 4.3 这样的临时变量声明。

```
public protocol OCDeclaration: OCAST {}

class OCVariableDeclaration: OCDeclaration {
    let variable: OCVar
    let type: String
    let right: OCAST
    
    init(variable: OCVar, type: String, right: OCAST) {
        self.variable = variable
        self.type = type
        self.right = right
    }
}
```

OCVariableDeclaration 节点的结构和赋值语句节点的结构类似，只是多了一个类型属性。接下来，在 OCParser 和 OCInterpreter 里对 OCVariableDeclaration 节点进行处理。

```
private func statement() -> OCAST {
    switch currentTk {
    case .id:
        if case .id = nextTk {
            guard case let .id(name) = currentTk else {
                fatalError("Error: wrong")
            }
            eat(.id(name))
            let v = variable()
            if currentTk == .assign {
                eat(.assign)
                let right = expr()
                return OCVariableDeclaration(variable: v, type: name, right: right)
            } else {
                fatalError("Error: wrong")
            }
        }
        return assignStatement()
    default:
        return empty()
    }
}
```

解释处理时也和赋值语句节点一样，将 = 右侧表达式的运算结果和声明的变量对应上即可。

```
func eval(variableDeclaration: OCVariableDeclaration) -> OCValue {
    scopes[variableDeclaration.variable.name] = eval(node: variableDeclaration.right)
    return .none
}
```

前面都是针对一个作用域进行的解释处理。如果是对多级作用域进行解释处理，就需要设计一个符号表。符号表需要包含当前符号所在作用域的层级，用 level 属性表示，类型为整型。还需要包含上级作用域，用 enclosingScope 属性表示，类型是 OCSymbolTable 可选类型。由于当前的作用域可能是最高一级的，那么 enclosingScope 属性就有可能为空，所以 enclosingScope 属性的类型后面需要加上一个问号，表示 enclosingScope 属性是可选类型。

```
public class OCSymbolTable {
    var symbols: [String: OCSymbol] = [:]

    let name: String
    let level: Int
    let enclosingScope: OCSymbolTable?

    init(name: String, level: Int, enclosingScope: OCSymbolTable?) {
        self.name = name
        self.level = level
        self.enclosingScope = enclosingScope
```

```
        defineBuiltInTypes()
    }

    private func defineBuiltInTypes() {
        define(OCBuiltInTypeSymbol.integer)
        define(OCBuiltInTypeSymbol.float)
        define(OCBuiltInTypeSymbol.boolean)
        define(OCBuiltInTypeSymbol.string)
    }

    func define(_ symbol: OCSymbol) {
        symbols[symbol.name] = symbol
    }

    func lookup(_ name: String, currentScopeOnly: Bool = false) -> OCSymbol? {
        if let symbol = symbols[name] {
            return symbol
        }
        if currentScopeOnly {
            return nil
        }
        return enclosingScope?.lookup(name)
    }
}
```

有了 enclosingScope 属性，在当前作用域找不到符号时，可以递归向上看上级作用域是否有，也可以通过 lookup 方法的参数 currentScopeOnly 来控制是否只在当前作用域里查找。

在 OCStaticAnalyzer 静态分析里，添加一个映射表记录各个作用域的符号表，同时添加一个 currentScope 记录当前访问的作用域对应的符号表。

```
public class OCStaticAnalyzer: OCVisitor {
    private var currentScope: OCSymbolTable?
    private var scopes: [String: OCSymbolTable] = [:]

    public init() {

    }

    public func analyze(node: OCAST) -> [String: OCSymbolTable] {
        visit(node: node)
        return scopes
    }

    func visit(program: OCProgram) {
        let globalScope = OCSymbolTable(name: "global", level: 1, enclosingScope: nil)
        scopes[globalScope.name] = globalScope
        currentScope = globalScope
        visit(interface: program.interface)
        visit(implementation: program.implementation)
        currentScope = nil
    }
```

```swift
        func visit(variableDeclaration: OCVariableDeclaration) {
            guard let scope = currentScope else {
                fatalError("Error: out of a scope")
            }
            
            guard scope.lookup(variableDeclaration.variable.name, currentScopeOnly: true) == nil else {
                fatalError("Error: Doplicate identifier")
            }
            
            guard let symbolType = scope.lookup(variableDeclaration.type)
                else {
                fatalError("Error: type not found")
            }
            
            scope.define(OCVariableSymbol(name: variableDeclaration.variable.name, type: symbolType))
            visit(node: variableDeclaration.variable)
            visit(node: variableDeclaration.right)
        }
        
        func visit(method: OCMethod) {
            let scope = OCSymbolTable(name: method.methodName, level: (currentScope?.level ?? 0) + 1, enclosingScope: currentScope)
            scopes[scope.name] = scope
            currentScope = scope
            
            for statement in method.statements {
                visit(node: statement)
            }
            
            currentScope = currentScope?.enclosingScope
        }
        
        func visit(propertyDeclaration: OCPropertyDeclaration) {
            guard let scope = currentScope else {
                fatalError("Error: out of a scope")
            }
            guard scope.lookup(propertyDeclaration.name) == nil else {
                fatalError("Error: duplicate identifier \(propertyDeclaration.name) found")
            }
            
            guard let symbolType = scope.lookup(propertyDeclaration.type) else {
                fatalError("Error: \(propertyDeclaration.type) type not found")
            }
            
            scope.define(OCVariableSymbol(name: propertyDeclaration.name, type: symbolType))
        }
        
        func visit(variable: OCVar) {
            guard let scope = currentScope else {
```

```
            fatalError("Error: cannot access")
        }
        guard scope.lookup(variable.name) != nil else {
            fatalError("Error: \(variable.name) variable not found")
        }
    }
}
```

执行静态分析，代码如下：

```
let sa = OCStaticAnalyzer()
let symtb = sa.analyze(node: ast)
```

从断点处可以看到，symtb 的结果是符合预期的。

```
▼ 🅛 symtb = ([String : HTN.OCSymbolTable]) 2 key/value pairs
  ▼ [0] ((key: String, value: HTN.OCSymbolTable))
    ▶ key = (String) "run"
    ▼ value = (HTN.OCSymbolTable) 0x000060c000069bc0
      ▶ symbols = ([String : OCSymbol]) 5 key/value pairs
      ▶ name = (String) "run"
        level = (Int) 2
      ▼ enclosingScope = (HTN.OCSymbolTable?) 0x000060c000069840
        ▶ symbols = ([String : OCSymbol]) 6 key/value pairs
        ▶ name = (String) "global"
          level = (Int) 1
          enclosingScope = (HTN.OCSymbolTable?) nil
  ▼ [1] ((key: String, value: HTN.OCSymbolTable))
    ▶ key = (String) "global"
    ▼ value = (HTN.OCSymbolTable) 0x000060c000069840
      ▶ symbols = ([String : OCSymbol]) 6 key/value pairs
      ▶ name = (String) "global"
        level = (Int) 1
        enclosingScope = (HTN.OCSymbolTable?) nil
```

Objective-C 的例子里有两个作用域，分别是 global 和 run。其中，run 的上级作用域是 global。加上 BuiltInTypeSymbol 的内置类型符号，symbols 的数量也能够对应上。

到此，我们已经写出了一个简单的解释器。作为基础教学，因为篇幅限制，在本章中我们并未实现一个语言所有必要的构造，比如递归、数组、字典、字符串、自定义数据结构、循环和条件表达式，等等。不过大家可以在第 2 章中学习更多的知识。

第 2 章
编译器

2.1 LLVM 简介

iOS 的开发语言 Objective-C 和 Swift 都是使用 LLVM 进行编译的。LLVM 是由 Chris Lattner 开发的，起初叫 "Low Level Virtual Machine"，后来因为使用的范围越来越广，可以用于常规编译器、JIT 编译器、汇编器、调试器、静态分析工具等一系列跟编程语言相关的工具开发，所以后来就被简称为 "LLVM"。Chris Lattner 后来又开发了 clang，使得 LLVM 直接挑战了 GCC 的地位。2012 年，LLVM 获得美国计算机学会 ACM 的软件系统大奖，从此和 UNIX、WWW、TCP/IP、Tex、Java 等齐名。Chris Lattner 加入苹果公司后，将苹果公司使用的 GCC 替换为 LLVM。2010 年，Chris Lattner 开始主导开发 Swift 语言。

LLVM 是工具链技术与一个模块化和可重用的编译器的集合。clang 是 LLVM 的子项目，是 C、C++ 和 Objective-C 的编译器，用于进行快速编译，其编译速度比 GCC 快 3 倍。其中 clang static analyzer 主要用于进行语法分析、语义分析和生成中间代码。在这个过程中会对代码进行检查，出错的代码和需要注意的代码会被标注出来。LLVM 核心库提供了一个优化器，对流行的 CPU 做代码生成支持。lld 是 clang / LLVM 的内置链接器，clang 必须调用链接器才能生成可执行文件。

LLVM 框架是围绕着代码编写良好的中间表示而构建的，所以 LLVM 比较有特色的一点就是它能提供一种代码编写良好的中间表示，这意味着它可以作为多种语言的后端，提供与语言无关的优化，同时能够针对多种 CPU 生成代码。

LLVM 还被用于在 Gallium3D 中进行 JIT 优化，Xorg 中的 Pixman 也考虑使用 LLVM 优化执行速度。LLVM-Lua 用 LLVM 来编译 Lua 代码。GPU Ocelot 使用 LLVM 来让 CUDA 程序无须重新编译就能够运行于装有不同型号 CPU 的设备。

2.2 编译流程

在列出完整步骤之前我们先看一个简单的例子，看看 LLVM 是如何完成编译的。

```
#import <Foundation/Foundation.h>
#define DEFINEEight 8

int main(){
    @autoreleasepool {
        int eight = DEFINEEight;
```

```
        int six = 6;
        NSString* site = [[NSString alloc] initWithUTF8String:"starming"];
        int rank = eight + six;
        NSLog(@"%@ rank %d", site, rank);
    }
    return 0;
}
```

在命令行输入以下命令。

```
clang -ccc-print-phases main.m
```

可以看到编译源文件需要经历的几个不同阶段。

```
0: input, "main.m", objective-c
1: preprocessor, {0}, objective-c-cpp-output
2: compiler, {1}, ir
3: backend, {2}, assembler
4: assembler, {3}, object
5: linker, {4}, image
6: bind-arch, "x86_64", {5}, image
```

这样能够了解整个过程中的重要信息。查看 Objective-C 的 C 语言实现源代码可以使用如下命令。

```
clang -rewrite-objc main.m
```

查看 clang 的内部命令，可以使用 -### 命令。

```
clang -### main.m -o main
```

想看清 clang 的全部过程，可以通过-E 查看 clang 在预编译处理时做了哪些工作。

```
clang -E main.m
```

执行完上面代码后，会在控制台输出如下内容。

```
# 1 "/System/Library/Frameworks/Foundation.framework/Headers
    /FoundationLegacySwiftCompatibility.h" 1 3
# 185 "/System/Library/Frameworks/Foundation.framework/Headers
    /Foundation.h" 2 3
# 2 "main.m" 2

int main(){
    @autoreleasepool {
        int eight = 8;
        int six = 6;
        NSString* site = [[NSString alloc] initWithUTF8String:"starming"];
        int rank = eight + six;
        NSLog(@"%@ rank %d", site, rank);
    }
    return 0;
}
```

预编译包括宏的替换、头文件的导入。下面这些代码也会在预编译里进行处理。

- "#define"

- "#include"
- "#indef"
- "#pragma"

预处理完成后，clang 就会进行词法分析。这里会把代码切成一个一个 token，比如大小括号、等于号和字符串等。

```
clang -fmodules -fsyntax-only -Xclang -dump-tokens main.m
```

然后进行语法分析，验证语法是否正确，再将所有节点组成抽象语法树。

```
clang -fmodules -fsyntax-only -Xclang -ast-dump main.m
```

完成这些步骤后就可以开始进行 IR 代码的生成了。CodeGen 会负责将抽象语法树自上而下遍历，逐步翻译成 LLVM IR。IR 代码既是前端编译的输出结果，又是后端编译的输入内容。

```
clang -S -fobjc-arc -emit-llvm main.m -o main.ll
```

在这里 LLVM 会做些优化的工作，在 Xcode 的编译设置里也可以设置优化级别 -O1、-O3、-Os，还可以写些自己的 Pass。

```
clang -O3 -S -fobjc-arc -emit-llvm main.m -o main.ll
```

Pass 是 LLVM 优化工作的一个节点。每个节点都做些优化的工作，所有节点一起完成了 LLVM 的所有优化和转化工作。

如果开启了 Bitcode，LLVM 会对代码做进一步的优化，这份优化生成的 Bitcode 可以用在后端架构中。

```
clang -emit-llvm -c main.m -o main.bc
```

生成汇编：

```
clang -S -fobjc-arc main.m -o main.s
```

生成目标文件：

```
clang -fmodules -c main.m -o main.o
```

生成可执行文件，这样就能够执行并看到输出结果：

```
clang main.o -o main
```

执行：

```
./main
```

输出：

```
starming rank 14
```

下面是完整步骤：

- 将编译信息写入辅助文件，创建文件架构 .app 文件。

- 处理文件的打包信息。
- 执行 CocoaPod 编译前脚本，checkPods Manifest.lock。
- 编译 .m 文件，使用 CompileC 和 clang 命令。
- 链接器会去链接程序所需要的 Framework。
- 编译 xib。
- 拷贝资源文件。
- 编译 ImageAsset。
- 处理 info.plist。
- 执行 CocoaPod 脚本。
- 拷贝标准库。
- 创建 .app 文件和签名。

2.3　使用 clang 命令编译 .m 文件

使用 Xcode 编译过后，我们可以在 Show the report navigator 里 Target 对应的构建日志中查看每个 .m 文件的 clang 命令的参数信息，这些参数都是通过 Build Setting 选项设置的。以 AFSecurityPolicy.m 的信息为例，首先对任务进行描述。

```
CompileC DerivedData path/AFSecurityPolicy.o AFNetworking/AFNetworking
/AFSecurityPolicy.m normal x86_64 objective-c com.apple.compilers.llvm
.clang.1_0.compiler
```

接下来，更新工作路径，同时设置 PATH。

```
cd /Users/didi/Documents/Demo/GitHub/GCDFetchFeed/GCDFetchFeed/Pods
    export LANG=en_US.US-ASCII
    export PATH="/Applications/Xcode.app/Contents/Developer/Platforms
            /iPhoneSimulator.platform/Developer/usr/bin:
            /Applications/Xcode.app/Contents/Developer/usr/bin:/usr
            /local/bin:/usr/bin:/bin:/usr/sbin:/sbin"
```

然后开始编译命令。

```
clang -x objective-c -arch x86_64 -fmessage-length=0 -fobjc-arc…
-Wno-missing-field-initializers … -DDEBUG=1 … -isysroot
iPhoneSimulator10.1.sdk -fasm-blocks … -I -F -c AFSecurityPolicy.m -o
AFSecurityPolicy.o
```

clang 命令参数说明：
- -x：指定后续输入文件的编译语言，比如 Objective-C。
- -arch：指定编译的架构，比如 ARM7。
- -f：以-f开头的命令参数，用来诊断、分析代码。
- -W：以-W 开头的命令参数，可以通过逗号分隔不同的参数以定制编译警告。
- -D：以-D 开头的命令参数，指的是预编译宏，通过这些宏可以实现条件编译。
- -I：添加目录到搜索路径中。

- -F：指需要的 Framework。
- -c：运行预处理、编译和汇编。
- -o：将编译结果输出到指定文件。

2.3.1 构建 Target

用编译工程中的第三方依赖库可以构建程序的 Target，按顺序输出如下信息。

```
Create product structure
Process product packaging
Run custom shell script 'Check Pods Manifest.lock'
Compile …
Link /Users/…
Copy …
Compile asset catalogs
Compile Storyboard file …
Process info.plist
Link Storyboards
Run custom shell script 'Embed Pods Frameworks'
Run custom shell script 'Copy Pods Resources'
…
Touch GCDFetchFeed.app
Sign GCDFetchFeed.app
```

从以上信息可以看出在各步骤中会分别调用不同的命令行工具进行执行。

2.3.2 Target 在构建过程中的控制

在 Xcode 的 Project Editor 中 Build Phases、Build Rules 和 Build Settings 选项控制编译的过程。

Build Phases

Build Phases 界面中的选项用于构建可执行文件的规则。用户可以指定 Target 的依赖项目，在 target build 之前需要先设置 Build 的依赖。在 Compile Sources 选项中指定所有必须编译的文件，这些文件会根据 Build Settings 和 Build Rules 选项里的设置进行处理。

在 Link Binary With Libraries 选项里会列出所有的静态库和动态库，它们会和编译生成的目标文件进行链接。

Build Phases 还具有把静态资源拷贝到 bundle 里的功能。

通过在 Build Phases 里添加自定义脚本可以修改工程配置信息，比如 CocoaPod。

Build Rules

Build Rules 界面中的选项用于指定不同类型的文件是如何编译的。Build Rules 指定了不同类型文件的处理方式，以及输出位置，还可以增加新规则指定特定文件类型的处理方法。

Build Settings

在构建的过程中对各个阶段的选项进行设置。

对构建过程的控制设置都会保存在工程文件 .pbxproj 里，在这个文件中可以找到 rootObject 的 id。

```
rootObject = 3EE311301C4E1F0800103FA3 /* Project object */;
```

根据这个 id 可以找到 main 工程的定义。

```
/* Begin PBXProject section */
      3EE311301C4E1F0800103FA3 /* Project object */ = {
            isa = PBXProject;
            …
/* End PBXProject section */
```

targets 代码段显示了工程里的 Target id，比如下面代码里的 GCDFetchFeed、GCDFetchFeedTests、GCDFetchFeedUITests。

```
targets = (
    3EE311371C4E1F0800103FA3 /* GCDFetchFeed */,
    3EE311501C4E1F0800103FA3 /* GCDFetchFeedTests */,
    3EE3115B1C4E1F0800103FA3 /* GCDFetchFeedUITests */,
);
```

通过这些 id 就能够找到更详细的定义。比如，通过 GCDFetchFeed 这个 Target 的 id 可以找到如下定义：

```
3EE311371C4E1F0800103FA3 /* GCDFetchFeed */ = {
    isa = PBXNativeTarget;
    buildConfigurationList = 3EE311651C4E1F0800103FA3 /* configuration list for PBXNativeTarget "GCDFetchFeed" */;
    buildPhases = (
        9527AA01F4AAE11E18397E0C /* Check Pods st.lock */,
        3EE311341C4E1F0800103FA3 /* Sources */,
        3EE311351C4E1F0800103FA3 /* Frameworks */,
        3EE311361C4E1F0800103FA3 /* Resources */,
        C3DDA7C46C0308459A18B7D9 /* Embed Pods Frameworks
        DD33A716222617FAB49F1472 /* Copy Pods Resources
    );
    buildRules = (
    );
    dependencies = (
    );
    name = GCDFetchFeed;
    productName = GCDFetchFeed;
    productReference = 3EE311381C4E1F0800103FA3 /* chFeed.app */;
    productType = "com.apple.product-type.application";
};
```

通过以上代码可以看到，buildConfigurationList 指向了可用的配置项，包含 Debug 和 Release。而且，通过 id 还能找到有关 buildPhases、buildRules 和 dependencies 更详细的定义。

2.4　clang static analyzer

clang static analyzer 是一个静态代码分析工具，可用于查找 C 语言、C++和 Objective-C

程序中的 bug。此工具在进行静态分析之前会将源代码拆分成多个 token，这个过程称为词法分析（Lexical Analysis）。在 LLDB 的 TokensKind.def 文件里可以查看 clang 定义的所有 token。token 可以分为以下几类：

- 关键字，指的是语法中的关键字 if、else、while、for 等。
- 标识符，指的是变量名。
- 字面量，包括值、数字、字符串等。
- 特殊符号，包括加、减、乘、除等符号。

通过下面的命令可以输出所有 token 和所在文件的具体位置。

```
clang -fmodules -E -Xclang -dump-tokens main.m
```

通过此命令可以获得每个 token 的类型、值和具体位置。

在完成词法分析后，clang static analyzer 接着进行语法分析，将 token 按照语法、语义规则组合成类似 VarDecl 的节点，然后将这些节点按照层级关系构成语法树。

打个比方，如果遇到 token 带有等号，则进行赋值处理，如果遇到加号、减号、乘号、除号，则先处理乘、除，然后处理加、减。这些组合经过嵌套后会生成一个语法树的结构，整个过程完成后，在赋值操作时判断类型是否匹配。

打印语法树的命令如下。

```
clang -fmodules -fsyntax-only -Xclang -ast-dump main.m
```

命令输出的结果中，TranslationUnitDecl 是根节点，表示一个源文件；Decl 表示一个声明；Expr 表示表达式；Literal 表示字面量是特殊的 Expr；Stmt 表示语句。

clang static analyzer 包括 analyzer core 和 checker 两部分，可以通过 clang-analyze 命令调用。所有 checker 都是基于底层 analyzer core 之上的，通过 analyzer core 提供的功能能够编写新的 checker。

可以通过 clang-analyze-Xclang-analyzer-checker-help 列出当前 clang 版本下的所有 checker。如果想编写自己的 checker，可以在 clang 项目的 lib / StaticAnalyzer / Checkers 目录下找到实例参考，比如通过 ObjCUnusedIVarsChecker.cpp 检查定义过但未使用的变量。这种方式能够方便用户检查代码规则，或者对 bug 类型进行扩展。但是这种方式也有不足，每执行完一条语句后，analyzer core 会遍历所有 checker 中的回调函数，所以 checker 越多，语句执行速度越慢。通过 clang -cc1 -analyzer-checker-help 可以列出能调用的 checker。下面是常用的 checker。

debug.ConfigDumper	Dump config table
debug.DumpCFG	Display Control-Flow Graphs
debug.DumpCallGraph	Display Call Graph
debug.DumpCalls	Print calls as they are traversed by the engine
debug.DumpDominators	Print the dominance tree for a given CFG

debug.DumpLiveVars	Print results of live variable analysis
debug.DumpTraversal	Print branch conditions as they are traversed by the engine
debug.ExprInspection	Check the analyzer's understanding of expressions
debug.Stats	Emit warnings with analyzer statistics
debug.TaintTest	Mark tainted symbols as such
debug.ViewCFG	View Control-Flow Graphs using GraphViz
debug.ViewCallGraph	View Call Graph using GraphViz
debug.ViewExplodedGraph	View Exploded Graphs using GraphViz

最常用的 checker 是 DumpCFG、DumpCallGraph、DumpLiveVars 和 ViewExplodedGraph。CheckerManager 通过 runCheckersOn*() 管理插件。

- runCheckersOnASTDecl()：遍历 checker 列表，找到 DeclKind 的 checker，然后运行回调函数。
- runCheckersOnASTBody()：分析函数体。
- runCheckersOnEndOfTranslationUnit()：分析编译单元，再调用回调函数。

clang static analyzer 包括 CFG、MemRegion、SValBuilder、ConstraintManager 和 ExplodedGraph 等模块。clang static analyzer 本质上就是 path-sensitive analysis，要更好地理解 clang static analyzer，就需要对 data flow analysis 有所理解。data flow analysis 包括 inline 式的路径敏感分析和非 inline 式的路径敏感分析，如下图所示。

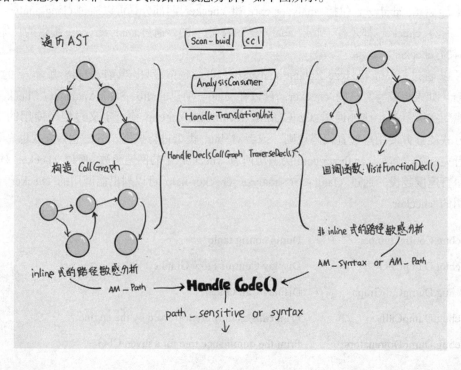

clang static analyzer 提供了很多辅助方法，比如 SVal.dump()、MemRegion.getString，以及 Stmt 和 Decl 提供的 Dump 方法。clang 语法树常见的 API 有 Stmt、Decl、Expr 和 QualType。在编写 checker 时会遇到语法树的层级检查，这时有个很好的接口——StmtVisitor 可以使用，这个接口具有类似 RecursiveASTVisitor 的作用。

整个 clang static analyzer 的入口是 AnalysisConsumer。回调方法 HandleTranslationUnit() 进行语法树层级分析或者进行 path-sensitive 分析。默认使用 inline 的 path-sensitive 分析，构建 CallGraph，从顶层 caller 按照调用的关系来分析，并且使用的是 WorkList 算法，从 EntryBlock 开始一步步模拟。这个过程被称为 Intra-Procedural Analysis（IPA）。这个模拟过程还需要对内存进行模拟，在 clang 的具体实现代码中可以查看这两个文件：MemRegion.h 和 RegionStore.cpp。

下面举个简单例子，看看 clang static analyzer 是如何对源代码进行模拟的。

```
int main()
{
    int a;
    int b = 10;
    a = b;
    return a;
}
```

对应的语法树以及 CFG 如下所示。

```
#————————AST————————
# clang -cc1 -ast-dump
TranslationUnitDecl 0xc75b450 <<invalid sloc>> <invalid sloc>
|-TypedefDecl 0xc75b740 <<invalid sloc>> <invalid sloc> implicit
__builtin_va_list 'char *'
 '-FunctionDecl 0xc75b7b0 <test.cpp:1:1, line:7:1> line:1:5 main 'int (void)'
   '-CompoundStmt 0xc75b978 <line:2:1, line:7:1>
     |-DeclStmt 0xc75b870 <line:3:2, col:7>
     | '-VarDecl 0xc75b840 <col:2, col:6> col:6 used a 'int'
     |-DeclStmt 0xc75b8d8 <line:4:2, col:12>
     | '-VarDecl 0xc75b890 <col:2, col:10> col:6 used b 'int' cinit
     |   '-IntegerLiteral 0xc75b8c0 <col:10> 'int' 10

<<<<<<<<<<<<<<<<<<<<<<<<<<<< a = b <<<<<<<<<<<<<<<<<<<<<<<<<<<<
     |-BinaryOperator 0xc75b928 <line:5:2, col:6> 'int' lvalue '='
     | |-DeclRefExpr 0xc75b8e8 <col:2> 'int' lvalue Var 0xc75b840 'a' 'int'
     | '-ImplicitCastExpr 0xc75b918 <col:6> 'int' <LValueToRValue>
     |   '-DeclRefExpr 0xc75b900 <col:6> 'int' lvalue Var 0xc75b890 'b' 'int'
<<<<<<<<<<<<<<<<<<<<<<<<<<<<<<<<<<<<<<<<<<<<<<<<<<<<<<<<<<<<<<<<

     '-ReturnStmt 0xc75b968 <line:6:2, col:9>
       '-ImplicitCastExpr 0xc75b958 <col:9> 'int' <LValueToRValue>
         '-DeclRefExpr 0xc75b940 <col:9> 'int' lvalue Var 0xc75b840 'a' 'int'
#————————CFG————————
# clang -cc1 -analyze -analyzer-checker=debug.DumpCFG
int main()
 [B2 (ENTRY)]
   Succs (1): B1
```

```
[B1]
 1: int a;
 2: 10
 3: int b = 10;
 4: b
 5: [B1.4] (ImplicitCastExpr, LValueToRValue, int)
 6: a
 7: [B1.6] = [B1.5]
 8: a
 9: [B1.8] (ImplicitCastExpr, LValueToRValue, int)
10: return [B1.9];
 Preds (1): B2
 Succs (1): B0

[B0 (EXIT)]
 Preds (1): B1
```

CFG 将程序拆分得更细，能够将执行的过程表现得更直观。为了避免"路径爆炸"，函数 inline 的条件会设置得更严格。当 CFG 块多时不会进行 inline 分析，当模拟栈深度超过一定值（默认值为 5）时，也不会进行 inline 分析。下面列出函数不会被 inline 的情况：

- AnalyzerOpts 设置项 inline 为否。
- 临时对象被析构函数调用。
- 合成函数。
- CFG 创建失败。
- BaseBlock 大于 50 个。
- inline 栈存在递归。
- inline 栈深度大于 5 层，并且 BaseBlock 不大于 3 个。
- BaseBlock 大于 14 个，inline 次数超过 32 次。

下图形象地表现了函数 inline 的过程：

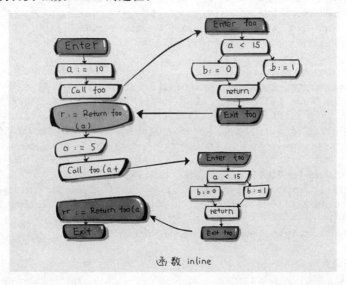

函数 inline

不能用函数 inline 怎么办呢？答案是使用 ConservativeEvalCall 函数。

在 MRC 中，执行路径使用的是 CFG 的执行路径模拟，在 ARC 中执行路径就没有使用 CFG 的执行路径模拟。举个例子，当全部条件没有都返回时，CFG 就会报错，而语法树就不会。

和 CFG 相关的类如下：
- CFGElement：最基本元素的基类。
- CFGTerminator：特殊的类，存在于 CFGBlock 之外。
- CFGBlock：由一系列 CFGElement 组成，包括 Predecessor CFGBlock、Successor CFGBlock 和 Terminator。
- CFG：由多个 CFGBlock 组成。

CFGElement 子类有：
- CFGStmt：Stmt 类型的语句。
- CFGInitializer：初始化语句。
- CFGNewAllocator：new 表达式。
- CFGImplicitDtor：隐式产生的析构函数。

CFGElement 子类的完整表示如下图所示。

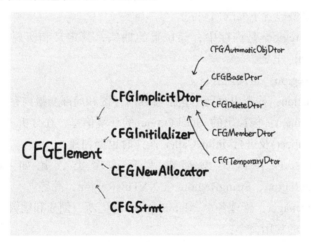

下面介绍 clang static analyzer 的一些重要类。
- ConstraintManager.h：最重要的类，用来管理 SymExpr 的限制条件。
- AnalysisManager.h：含有核心内容，比如 CheckerManager、DiagnosticsEngine，以及 ASTContext。
- APSIntType.h：用于在进行静态分析时表示各种不同类型的 Integer。
- BasicValueFactory.h：引擎的整数池，存放静态分析过程中得到的 APSInt，处理整形边界以及类型。比如，malloc(-1)会分配大块内存区域，while(char < 300)会导致死循环。
- CoreEngine.h：在 CFG 上进行符号执行，用于构建 ExplodedGraph。
- ExprEngine.h：构建在 CoreEngine 上。
- SVals.h：定义 SVal、Loc 和 NonLoc 等类，用于表达静态分析中获得的值，比如变量值和内存地址值。
- SValBuilder.h：创建符号执行。
- Environment.h：存储当前能引用的表达式与对应符号值之间的映射关系，生命周期只限于当前语句，比如 if(foo(a) + 1 > b) {}，foo(a) -> SVal，b -> SVal。
- ExplodedGraph.h：在 CFG 上进行符号执行的结果。
- ProgramState.h：静态分析过程中的程序状态。
- BlockCounter.h：记录一条路径上某个块共执行了多少次。
- FunctionSummary.h：记录了函数的信息，在 inline 时有非常重要的作用。
- CallEvent.h：定义了函数调用。
- Store.h：对存储进行修改。
- StoreRef.h：Store 的包装类。
- MemRegion.h：对内存位置进行抽象的类，包括 MemRegion、SubRegion、AllocaRegion、MemSpaceRegion 和 StackRegion。
- CheckerContext.h：给 checker 提供上下文信息。
- WorkList.h：工作队列算法，DFS、BFS 和 BFSBlockDFSContents 三个类继承自它，表示不同的 CFG 遍历规则。

静态分析内存模型包含下面的模块：

- MemRegion：region 是内存中存储位置的抽象，其定义和实现在 MemRegion.h 和 MemRegion.cpp 里。
- GlobalSpaceRegion：全局内存区。
- StackSpaceRegion：栈上内存区，包括参数内存区和局部变量内存区。
- StackFrameContext：模拟出的 StackFrame 的环境信息，在分析 FunctionDecl 时会创建，在分析 callee()及进行 inlineCall() 操作时也会创建。
- TypedRegion：内存区类型都继承自这个类，比如 CodeTextRegion、FunctionCodeRegion、StringRegion、CXXThisRegion，等等。
- MemRegionManager：管理各个 MemRegion 的更新、创建和销毁。

下面是用于调试的常用命令：

- clang-cc1-analyzer-checker-help：打印可用 checker。
- clang-cc1-analyze-analyzer-checker=debug.DumpCFG main.c：CFG 信息。
- clang-cc1 -analyze-analyzer-checker=debug.ViewExplodedGraph main.c：Exploded Graph 信息。
- clang -Xclang -ast-dump main.c -c：AST 信息。

2.5 IR 代码

LLVM 将语法树翻译成 IR 代码，作为后端输入的桥接语言，方便后端针对多种语言做相同的优化。

在这个过程中，LLVM 还会跟 runtime 桥接。桥接的类型包括以下几种：

- 各种类、方法、成员变量等的结构体的生成，并将其放到对应的 Mach-O 的 section 中。
- Non-Fragile ABI 可以标注多个 OBJC_IVAR_$_ 偏移值常量。
- 将 ObjCMessageExpr 翻译成相应版本的 objc_msgSend，将 super 翻译成 objc_msgSendSuper。
- Strong、weak、copy、atomic 组合成 @property，自动实现 setter 和 getter。
- 对 @synthesize 的处理。
- 生成 block_layout 数据结构。
- __block 和 __weak 的处理。
- _block_invoke。
- ARC 处理包括插入 objc_storeStrong 和 objc_storeWeak 等 ARC 代码；ObjCAutoreleasePoolStmt 转 objc_autorealeasePoolPush / Pop；自动添加 [super dealloc]；给每个 ivar 的类合成 .cxx_destructor 方法，自动释放类的成员变量。

不管编译的是 Objective-C 还是 Swift 代码，也不管对应的硬件平台是什么类型，抑或是否为即时编译，LLVM 里唯一不变的是中间语言—— LLVM IR。下面我们就来看一看如何使用 LLVM IR。

2.5.1 IR 结构

下面是一个中间代码文件 main.ll 的示例。

```
; ModuleID = 'main.c'
source_filename = "main.c"
target datalayout = "e-m:o-i64:64-f80:128-n8:16:32:64-S128"
target triple = "x86_64-apple-macosx10.12.0"

@.str = private unnamed_addr constant [16 x i8] c"Please input a:\00", align 1
@.str.1 = private unnamed_addr constant [3 x i8] c"%d\00", align 1
@.str.2 = private unnamed_addr constant [16 x i8] c"Please input b:\00", align 1
@.str.3 = private unnamed_addr constant [32 x i8] c"a is:%d,b is :%d,count equal:%d\00", align 1

; Function Attrs: nounwind ssp uwtable
define i32 @main() #0 {
  %1 = alloca i32, align 4
  %2 = alloca i32, align 4
  %3 = bitcast i32* %1 to i8*
  call void @llvm.lifetime.start(i64 4, i8* %3) #3
  %4 = bitcast i32* %2 to i8*
```

```
        call void @llvm.lifetime.start(i64 4, i8* %4) #3
    %5 = tail call i32 (i8*, ...) @printf(i8* getelementptr inbounds ([16 x i8], [16
x i8]* @.str, i64 0, i64 0))
    %6 = call i32 (i8*, ...) @scanf(i8* getelementptr inbounds ([3 x i8], [3 x i8]*
@.str.1, i64 0, i64 0), i32* nonnull %1)
    %7 = call i32 (i8*, ...) @printf(i8* getelementptr inbounds ([16 x i8], [16 x i8]*
@.str.2, i64 0, i64 0))
    %8 = call i32 (i8*, ...) @scanf(i8* getelementptr inbounds ([3 x i8], [3 x i8]*
@.str.1, i64 0, i64 0), i32* nonnull %2)
    %9 = load i32, i32* %1, align 4, !tbaa !2
    %10 = load i32, i32* %2, align 4, !tbaa !2
    %11 = add nsw i32 %10, %9
    %12 = call i32 (i8*, ...) @printf(i8* getelementptr inbounds ([32 x i8], [32 x i8]*
@.str.3, i64 0, i64 0), i32 %9, i32 %10, i32 %11)
    call void @llvm.lifetime.end(i64 4, i8* %4) #3
    call void @llvm.lifetime.end(i64 4, i8* %3) #3
    ret i32 0
}

; Function Attrs: argmemonly nounwind
declare void @llvm.lifetime.start(i64, i8* nocapture) #1

; Function Attrs: nounwind
declare i32 @printf(i8* nocapture readonly, ...) #2

; Function Attrs: nounwind
declare i32 @scanf(i8* nocapture readonly, ...) #2

; Function Attrs: argmemonly nounwind
declare void @llvm.lifetime.end(i64, i8* nocapture) #1

attributes #0 = { nounwind ssp uwtable "disable-tail-calls"="false"
"less-precise-fpmad"="false" "no-frame-pointer-elim"="true"
"no-frame-pointer-elim-non-leaf" "no-infs-fp-math"="false"
"no-nans-fp-math"="false" "stack-protector-buffer-size"="8"
"target-cpu"="penryn" "target-features"=
"+cx16,+fxsr,+mmx,+sse,+sse2,+sse3,+sse4.1,+ssse3"
"unsafe-fp-math"="false" "use-soft-float"="false" }
attributes #1 = { argmemonly nounwind }
attributes #2 = { nounwind "disable-tail-calls"="false"
"less-precise-fpmad"="false" "no-frame-pointer-elim"="true"
"no-frame-pointer-elim-non-leaf" "no-infs-fp-math"="false"
"no-nans-fp-math"="false" "stack-protector-buffer-size"="8"
"target-cpu"="penryn" "target-features"=
"+cx16,+fxsr,+mmx,+sse,+sse2,+sse3,+sse4.1,+ssse3"
"unsafe-fp-math"="false" "use-soft-float"="false" }
attributes #3 = { nounwind }

!llvm.module.flags = !{!0}
!llvm.ident = !{!1}

!0 = !{i32 1, !"PIC Level", i32 2}
!1 = !{!"Apple LLVM version 8.0.0 (clang-800.0.42.1)"}
!2 = !{!3, !3, i64 0}
```

```
!3 = !{!"int", !4, i64 0}
!4 = !{!"omnipotent char", !5, i64 0}
!5 = !{!"Simple C/C++ TBAA"}
```

以下是 IR 关键字的介绍：

- @：代表全局变量。
- %：代表局部变量。
- alloca：为当前执行的函数分配内存，当该函数执行完毕时自动释放内存。
- i32：表示整数占几位，例如 i32 就代表 32 位，即 4 字节。
- align：对齐。比如单个 int 占 4 字节，为了对齐，只占 1 字节的 char 要对齐，就需要占用 4 字节空间。
- load：读出。
- store 写入。
- icmp：两个整数值比较，返回布尔值。
- br：选择分支，根据条件跳转到对应的 label。
- label：代码标签。

IR 作为中间语言具有与语言无关的特性。下面是 IR 中与语言无关的类型信息：

- 语言共有的基本类型（void、bool、signed 等）。
- 复杂类型，pointer、array、structure 和 function。
- 弱类型的支持，用 cast 来实现一种类型到任意类型的转换。
- 支持地址运算，getelementptr 指令用于获取结构体子元素，比如 a.b 或 [a b]。

LLVM IR 有三种表示格式：第一种是像 Bitcode 这样的存储格式，以 .bc 做后缀；第二种是可读的 .ll；第三种是用于开发时操作 LLVM IR 的内存格式。

一个编译单元(即一个文件)在 IR 里就是一个 Module。Module 里有 Global Variable 和 Function，在 Function 里有 Basic Block。在 Basic Block 里有指令 Instruction。

通过下面的 IR 结构图，我们能够更好地理解 IR 的整体结构。

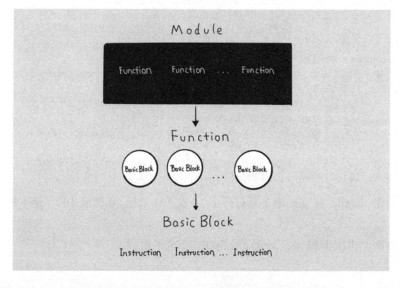

从上图可以看出层级最高的是 Module，其包含多个 Function。每个 Function 包含多个 BasicBlock，BasicBlock 含有 Instruction，结构非常清晰。如果想开发一个新语言，只需要在完成语法解析后，通过 LLVM 提供的丰富接口在内存中生成 IR，就可以直接在各个不同的平台上运行了。

IR 语言支持静态单赋值，可以很好地降低数据流分析和控制流分析的复杂度。但 IR 只能在定义时赋值，后面不能更改。这样就没法写程序了，编写输入、输出程序时会很困难，所以函数式编程才会有类似 Monad 这样的机制。Monad 的底层复杂度是非常高的，仅将输入、输出的状态通过类型系统重载，就可以把它们隐藏起来。

2.5.2　LLVM IR 优化

使用 O2、O3 这样的优化会调用对应的 Pass 来进行处理，比如进行死代码清理、内联化、表达式重组、循环变量移动，并且可以通过 llvm-opt 调用 LLVM 优化相关的库。

举个例子，我们看看 Pass 会做哪些优化。下面看一段 while 代码和转换成 IR 后对应的代码。

```
int i = 0;
while (i < 10) {
    i++;
    printf("%d",i);
}
```

对应的 IR 代码是：

```
  %call4 = call i32 (i8*, ...) @printf(i8* getelementptr inbounds ([3 x i8], [3 x i8]* @.str.1, i64 0, i64 0), i32 1)
  %call4.1 = call i32 (i8*, ...) @printf(i8* getelementptr inbounds ([3 x i8], [3 x i8]* @.str.1, i64 0, i64 0), i32 2)
  %call4.2 = call i32 (i8*, ...) @printf(i8* getelementptr inbounds ([3 x i8], [3 x i8]* @.str.1, i64 0, i64 0), i32 3)
  %call4.3 = call i32 (i8*, ...) @printf(i8* getelementptr inbounds ([3 x i8], [3 x i8]* @.str.1, i64 0, i64 0), i32 4)
  %call4.4 = call i32 (i8*, ...) @printf(i8* getelementptr inbounds ([3 x i8], [3 x i8]* @.str.1, i64 0, i64 0), i32 5)
  %call4.5 = call i32 (i8*, ...) @printf(i8* getelementptr inbounds ([3 x i8], [3 x i8]* @.str.1, i64 0, i64 0), i32 6)
  %call4.6 = call i32 (i8*, ...) @printf(i8* getelementptr inbounds ([3 x i8], [3 x i8]* @.str.1, i64 0, i64 0), i32 7)
  %call4.7 = call i32 (i8*, ...) @printf(i8* getelementptr inbounds ([3 x i8], [3 x i8]* @.str.1, i64 0, i64 0), i32 8)
  %call4.8 = call i32 (i8*, ...) @printf(i8* getelementptr inbounds ([3 x i8], [3 x i8]* @.str.1, i64 0, i64 0), i32 9)
  %call4.9 = call i32 (i8*, ...) @printf(i8* getelementptr inbounds ([3 x i8], [3 x i8]* @.str.1, i64 0, i64 0), i32 10)
```

可以看出来，while 在 IR 中重复打印了 10 次。如果把 10 改成 100，是不是会变成打印 100 次呢？

改成 100 后，再次生成 IR，可以看到 IR 变成了这样：

```
    br label %while.body

  while.body:                                 ; preds = %while.body, %entry
    %i.010 = phi i32 [ 0, %entry ], [ %inc, %while.body ]
    %inc = add nuw nsw i32 %i.010, 1
    %call4 = call i32 (i8*, ...) @printf(i8* getelementptr inbounds ([3 x i8], [3
 x i8]* @.str.1, i64 0, i64 0), i32 %inc)
    %exitcond = icmp eq i32 %inc, 100
    br i1 %exitcond, label %while.end, label %while.body

  while.end:                                  ; preds = %while.body
    %2 = load i32, i32* %a, align 4, !tbaa !2
    %3 = load i32, i32* %b, align 4, !tbaa !2
    %add = add nsw i32 %3, %2
    %call5 = call i32 (i8*, ...) @printf(i8* getelementptr inbounds ([11 x i8], [11
 x i8]* @.str.3, i64 0, i64 0), i32 %add)
    call void @llvm.lifetime.end(i64 4, i8* nonnull %1) #3
    call void @llvm.lifetime.end(i64 4, i8* nonnull %0) #3
    ret i32 0
  }
```

Pass 优化器根据不同条件生成不同的 IR 代码。br 会选择跳向 while.body 定义的标签。

```
br label %while.body
```

从这个标签可以看到：

```
%exitcond = icmp eq i32 %inc, 100
  br i1 %exitcond, label %while.end, label %while.body
```

icmp 会比较当前的 %inc 和定义的临界值 100，根据返回的布尔值是 true 还是 false 来决定 br 跳转到哪个代码标签，值是 true 就跳转到 while.end 标签，值是 false 就跳转到 while.body 标签。这就是 while 的逻辑。通过 br 跳转和 label 这种标签概念，使得 IR 语言能够更方便转成低级语言，同时具有更好的兼容性。

2.5.3 SSA

LLVM IR 是 SSA 形式的，维护双向 def-use 信息。use-def 是通过普通指针实现信息维护的，def-use 是通过内存跳表和链表来实现的，便于使用 forward dataflow analysis 和 backward dataflow analysis。可以通过 ADCE Pass 来了解一下 backward dataflow。ADCE Pass 的源文件在 lib/Transforms/Scalar/ADCE.cpp 中。ADCE Pass "乐观"地假设所有指令都是"死"的，直到证明是否定的，才允许消除其他 DCE Pass 的 Dead 捕获计算，特别是当涉及循环计算时。关于其他 DCE 相关的 Pass，可以查看同级目录下的 BDCE.cpp 和 DCE.cpp，目录下其他的 Pass 都是和数据流相关的分析，包含了各种分析算法和思路。

我们看一看加法操作的相关 IR 代码：

```
%2 = load i32, i32* %a, align 4, !tbaa !2
%3 = load i32, i32* %b, align 4, !tbaa !2
%add = add nsw i32 %3, %2
```

加法操作对应的指令是：

```
BinaryOperator::CreateAdd(Value *V1, Value *V2, const Twine &Name)
```

两个输入（V1 和 V2）的 def-use 是如何做的呢？可以参考如下代码：

```cpp
class Value {
  void addUse(Use &U) { U.addToList(&UseList); }
  // ...
};

class Use {
  Value *Val;
  Use *Next;
  PointerIntPair<Use **, 2, PrevPtrTag> Prev;
  // ...
};

void Use::set(Value *V) {
  if (Val) removeFromList();
  Val = V;
  if (V) V->addUse(*this);
}

Value *Use::operator=(Value *RHS) {
  set(RHS);
  return RHS;
}

class User : public Value {
  template <int Idx, typename U> static Use &OpFrom(const U *that) {
    return Idx < 0
      ? OperandTraits<U>::op_end(const_cast<U*>(that))[Idx]
      : OperandTraits<U>::op_begin(const_cast<U*>(that))[Idx];
  }
  template <int Idx> Use &Op() {
    return OpFrom<Idx>(this);
  }
  template <int Idx> const Use &Op() const {
    return OpFrom<Idx>(this);
  }
  // ...
};

class Instruction : public User,
                    public ilist_node_with_parent<Instruction, BasicBlock> {
  // ...
};

class BinaryOperator : public Instruction {
  /// Construct a binary instruction, given the opcode and the two
  /// operands. Optionally (if InstBefore is specified) insert the instruction
  /// into a BasicBlock right before the specified instruction. The specified
  /// Instruction is allowed to be a dereferenced end iterator.
  ///
  static BinaryOperator *Create(BinaryOps Op, Value *S1, Value *S2,
```

```cpp
                    const Twine &Name = Twine(),
                    Instruction *InsertBefore = nullptr);
  // ...
};

BinaryOperator::BinaryOperator(BinaryOps iType, Value *S1, Value *S2,
                   Type *Ty, const Twine &Name,
                   Instruction *InsertBefore)
  : Instruction(Ty, iType,
            OperandTraits<BinaryOperator>::op_begin(this),
            OperandTraits<BinaryOperator>::operands(this),
            InsertBefore) {
  Op<0>() = S1;
  Op<1>() = S2;
  init(iType);
  setName(Name);
}

BinaryOperator *BinaryOperator::Create(BinaryOps Op, Value *S1, Value *S2,
                   const Twine &Name,
                   Instruction *InsertBefore) {
  assert(S1->getType() == S2->getType() &&
      "Cannot create binary operator with two operands of differing type!");
  return new BinaryOperator(Op, S1, S2, S1->getType(), Name, InsertBefore);
}
```

从代码里可以看出，def-use 使用 Use 对象把 use 和 def 联系起来了。

LLVM IR 通过 mem2reg Pass 把局部变量变成 SSA 形式。这个 Pass 的代码在 lib/Transforms/Utils/Mem2Reg.cpp 里。LLVM 通过 mem2reg Pass 能够识别 alloca 模式，将其设置为 SSA value。这时就不再需要 alloca、load 和 store 了。mem2reg 是对 PromoteMemToReg 函数调用的一个简单包装，真正的算法实现在 PromoteMemToReg 函数里，该函数的说明在 lib/Transforms/Utils/ PromoteMemoryToRegister.cpp 文件里。

这个算法会使 alloca 仅仅作为 load 和 store 用途的指令，使用迭代 dominator 边界转换成 PHI 节点，然后使用深度优先函数排序重写 load 和 store。这种算法被称为 iterated dominance frontier，具体实现方法可以参看 PromoteMemToReg 函数的实现。

当然，把多个字节码 .bc 合成一个文件，链接时还会优化。IR 结构在优化后会有变化。我们可以在变化后的 IR 的结构上再进行更多的优化，并且可以进行 lli 解释，执行 LLVM IR。

llc 是专门编译 LLVM IR 的编译器，用以生成汇编文件。

LLVM 调用系统汇编器（比如 GNU 的 as）来编译生成 .o（Object）文件，然后用链接器链接相关库和.o 文件一起生成可执行的 .out 文件或者.exe 文件。

llvm-mc 可以直接生成.o 文件。

2.6 clang 前端组件

2.6.1 库的介绍

熟悉 clang 的库以后就能够自己动手编写程序了。下面介绍一下常用库，然后着重分析 libclang 库，并介绍如何用它编写程序。

- LLVM Support Library：提供了许多底层库和数据结构，包括命令行 option 处理、各种容器和系统抽象层，以及访问文件系统。
- Basic Library：提供了跟踪和操纵 source buffer，及其位置的能力，还提供了 diagnostic、token、抽象目标，以及编译语言子集信息的 low-level 实用程序。
- Driver Library：和 Driver 相关的库。
- Precompiled Header：clang 支持预编译 header 的实现。
- Frontend Library：这个库包含了在 clang 的库之上构建的功能，比如输出 diagnositic 等方法。
- Lexer and Preprocessor Library：词法分析和预处理库。
- Sema Library：解析器调用此库时，会对输入的代码进行语义分析。对于有效的程序，Sema Library 会为解析构造一个语法树。
- The CodeGen Library：此库用语法树作为输入，并以此生成 LLVM IR 代码。

2.6.2 使用 libclang 进行语法分析

使用 libclang 提供的方法可以对源文件进行语法分析，分析语法树，遍历语法树上每个节点。

使用 libclang 时可以直接使用 C 语言的 API，官方也提供了 Python Binding，还有开源的 node-js / ruby binding，以及 Objective-C 的开源库。

libclang 会让你觉得 clang 不仅仅是一个伟大的编译器，还是一个提供了强大静态检查功能的工具。下面我们从解析源代码开始介绍。

首先编写一个 libclang 程序来解析源代码：

```
int main(int argc, char *argv[]) {
    CXIndex Index = clang_createIndex(0, 0);
    CXTranslationUnit TU = clang_parseTranslationUnit(Index, 0,
                                            argv, argc, 0, 0,
                                            CXTranslationUnit_None);
    for (unsigned I = 0, N = clang_getNumDiagnostics(TU); I != N; ++I) {
        CXDiagnostic Diag = clang_getDiagnostic(TU, I);
        CXString String =
clang_formatDiagnostic(Diag,clang_defaultDiagnosticDisplayOptions());
        fprintf(stderr, "%s\n", clang_getCString(String));
        clang_disposeString(String);
    }
    clang_disposeTranslationUnit(TU);
    clang_disposeIndex(Index);
    return 0;
```

}

然后再编写一个有问题的 C 语言程序：

```
struct List { /**/ };
int sum(union List *L) { /* ... */ }
```

运行 libclang 程序进行语法检查后，会出现提示信息：

```
list.c:2:9: error: use of 'List' with tag type that does not match
    previous declaration
int sum(union List *Node) {
^~~~
struct
list.c:1:8: note: previous use is here
struct List {
^
```

最后我们分析诊断过程。下面显示了几个核心诊断方法，诊断出了问题。

- enum CXDiagnosticSeverity clang_getDiagnosticSeverity(CXDiagnostic Diag);
- CXSourceLocation clang_getDiagnosticLocation(CXDiagnostic Diag);
- CXString clang_getDiagnosticSpelling(CXDiagnostic Diag);

对有问题的地方进行高亮显示，并且提供两个修复方法。

- unsigned clang_getDiagnosticNumFixIts(CXDiagnostic Diag);
- CXString clang_getDiagnosticFixIt(CXDiagnostic Diag, unsigned FixIt, CXSourceRange *ReplacementRange);

遍历语法树的节点。C 语言源程序如下：

```
struct List {
    int Data;
    struct List *Next;
};
int sum(struct List *Node) {
    int result = 0;
    for (; Node; Node = Node->Next)
        result = result + Node->Data;
    return result;
}
```

首先找出所有的声明，比如 List、Data、Next、Sum、Node，以及 Result 等。然后找出引用，比如 struct List *Next 里的 List，以及声明和表达式，比如 int result = 0、 for 语句，以及宏定义和实例化。

CXCursor 会统一语法树的节点，规范包含的信息如下：

- 代码所在位置和长度
- 名字和符号解析
- 类型
- 子节点

下面以 CXCursor 为例进行分析。

```
struct List {
    int Data;
    struct List *Next;
};
```

CXCursor 的处理过程如下:

```
//Top-level cursor C
clang_getCursorKind(C) == CXCursor_StructDecl
clang_getCursorSpelling(C) == "List" //获取名字字符串
clang_getCursorLocation(C)  //位置
clang_getCursorExtent(C)    //长度
clang_visitChildren(C, ...); //访问子节点

//Reference cursor R
clang_getCursorKind(R) == CXCursor_TypeRef
clang_getCursorSpelling(R) == "List"
clang_getCursorLocation(R)
clang_getCursorExtent(R)
clang_getCursorReferenced(R) == C //指向 C
```

2.7 Driver

接下来，我们来学习和 LLVM 交互的 Driver。

Driver 是 clang 面向用户的接口，用来解析 option 设置，决定调用哪些工具，以完成整个编译过程。

整个 Driver 源代码的入口函数就是 driver.cpp 里的 main 函数，它可以作为入口帮助我们分析整个 Driver 是如何工作的。

```
int main(int argc_, const char **argv_) {
  llvm::sys::PrintStackTraceOnErrorSignal(argv_[0]);
  llvm::PrettyStackTraceProgram X(argc_, argv_);
  llvm::llvm_shutdown_obj Y; // Call llvm_shutdown() on exit.

  if (llvm::sys::Process::FixupStandardFileDescriptors())
    return 1;

  SmallVector<const char *, 256> argv;
  llvm::SpecificBumpPtrAllocator<char> ArgAllocator;
  std::error_code EC = llvm::sys::Process::GetArgumentVector(
      argv, llvm::makeArrayRef(argv_, argc_), ArgAllocator);
  if (EC) {
    llvm::errs() << "error: couldn't get arguments: " << EC.message()
                 << '\n';
    return 1;
  }

  llvm::InitializeAllTargets();
  std::string ProgName = argv[0];
  std::pair<std::string, std::string> TargetAndMode =
      ToolChain::getTargetAndModeFromProgramName(ProgName);
```

```cpp
  llvm::BumpPtrAllocator A;
  llvm::StringSaver Saver(A);

  //省略
  ...

  // If we have multiple failing commands, we return the result of the first
  // failing command.
  return Res;
}
```

2.7.1　Driver 的工作流程

在 driver.cpp 的 main 函数里有 Driver 初始化的代码。我们来看一看和 Driver 相关的代码。

```cpp
  Driver TheDriver(Path, llvm::sys::getDefaultTargetTriple(), Diags);
  SetInstallDir(argv, TheDriver, CanonicalPrefixes);

  insertTargetAndModeArgs(TargetAndMode.first, TargetAndMode.second,
argv,SavedStrings);

  SetBackdoorDriverOutputsFromEnvVars(TheDriver);

  std::unique_ptr<Compilation> C(TheDriver.BuildCompilation(argv));
  int Res = 0;
  SmallVector<std::pair<int, const Command *>, 4> FailingCommands;
  if (C.get())
    Res = TheDriver.ExecuteCompilation(*C, FailingCommands);

  // Force a crash to test the diagnostics.
  if (::getenv("FORCE_CLANG_DIAGNOSTICS_CRASH")) {
    Diags.Report(diag::err_drv_force_crash) << "FORCE_CLANG_DIAGNOSTICS"
      "_CRASH";

    // Pretend that every command failed.
    FailingCommands.clear();
    for (const auto &J : C->getJobs())
      if (const Command *C = dyn_cast<Command>(&J))
        FailingCommands.push_back(std::make_pair(-1, C));
  }

  for (const auto &P : FailingCommands) {
    int CommandRes = P.first;
    const Command *FailingCommand = P.second;
    if (!Res)
      Res = CommandRes;

    // If result status is < 0, then the driver command signalled an error.
    // If result status is 70, then the driver command reported a fatal error.
    // On Windows, abort will return an exit code of 3. In these cases,
    // generate additional diagnostic information if possible.
    bool DiagnoseCrash = CommandRes < 0 || CommandRes == 70;
```

```
#ifdef LLVM_ON_WIN32
    DiagnoseCrash |= CommandRes == 3;
#endif
    if (DiagnoseCrash) {
      TheDriver.generateCompilationDiagnostics(*C, *FailingCommand);
      break;
    }
  }
```

可以看到初始化 Driver 后，Driver 会调用 BuildCompilation 生成 Compilation。Compilation 的字面意思是"合集"，通过 driver.cpp 的 include 可以看到如下信息。

```
#include "clang/Driver/Compilation.h"
```

根据此路径可以看到 Compilation 为 Driver 设置的一组任务的类。通过这个类，我们可以提取类里面 BuildCompilation 阶段比较关键的信息。

```
class Compilation {
  /// The original (untranslated) input argument list.
  llvm::opt::InputArgList *Args;

  /// The driver translated arguments. Note that toolchains may perform their
  /// own argument translation.
  llvm::opt::DerivedArgList *TranslatedArgs;
  /// The driver we were created by.
  const Driver &TheDriver;

  /// The default tool chain.
  const ToolChain &DefaultToolChain;
  ...
  /// The list of actions.
  /// This is maintained and modified by consumers, via
  /// getActions().
  ActionList Actions;

  /// The root list of jobs.
  JobList Jobs;
    ...
public:
    ...
  const Driver &getDriver() const { return TheDriver; }

  const ToolChain &getDefaultToolChain() const { return DefaultToolChain; }
    ...
  ActionList &getActions() { return Actions; }
  const ActionList &getActions() const { return Actions; }
    ...
  JobList &getJobs() { return Jobs; }
  const JobList &getJobs() const { return Jobs; }

  void addCommand(std::unique_ptr<Command> C) {
      Jobs.addJob(std::move(C));
  }
    ...
```

```
/// ExecuteCommand - Execute an actual command.
///
/// \param FailingCommand - For non-zero results, this will be set to the
/// Command which failed, if any.
/// \return The result code of the subprocess.
int ExecuteCommand(const Command &C, const Command *&FailingCommand) const;

/// ExecuteJob - Execute a single job.
///
/// \param FailingCommands - For non-zero results,
/// this will be a vector of
/// failing commands and their associated result code.
void ExecuteJobs(
    const JobList &Jobs,
    SmallVectorImpl<std::pair<int, const Command *>> &FailingCommands) const;
    ...
};
```

通过这些关键定义，再结合 BuildCompilation 函数的实现，可以看出 Driver 的流程是按照下图所示顺序完成的。

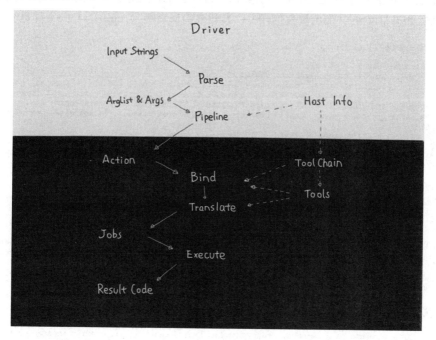

2.7.2 Parse

了解了完整的 Driver 流程后，我们接下来认识一下 Parse 。

Parse 是解析选项，对应的代码在 ParseArgStrings 这个函数里。

下面先执行一个 clang 命令试试，比如 clang -### main.c -ITheOptionWeAdd：

```
→ llvmdemo clang -### main.c -ITheOptionWeAdd
clang version 5.0.0 (trunk 294089)
Target: x86_64-apple-darwin16.4.0
Thread model: posix
InstalledDir: /usr/local/bin
 "/usr/local/bin/clang-5.0" "-cc1" "-triple" "x86_64-apple-macosx10.12.0" "-Wdeprecated-ob
jc-isa-usage" "-Werror=deprecated-objc-isa-usage" "-emit-obj" "-mrelax-all" "-disable-free
" "-main-file-name" "main.c" "-mrelocation-model" "pic" "-pic-level" "2" "-mthread-model"
 "posix" "-mdisable-fp-elim" "-masm-verbose" "-munwind-tables" "-target-cpu" "penryn" "-tar
get-linker-version" "274.2" "-dwarf-column-info" "-debugger-tuning=lldb" "-resource-dir"
 "/usr/local/bin/../lib/clang/5.0.0" "-I" "TheOptionWeAdd" "-fdebug-compilation-dir" "/Users
/didi/Downloads/llvmdemo" "-ferror-limit" "19" "-fmessage-length" "90" "-stack-protector"
 "1" "-fblocks" "-fobjc-runtime=macosx-10.12.0" "-fencode-extended-block-signature" "-fmax-
type-align=16" "-fdiagnostics-show-option" "-fcolor-diagnostics" "-o" "/var/folders/r9/35q
9g3d56_d9g0v59w9x2l9w0000gn/T/main-85975b.o" "-x" "c" "main.c"
 "/usr/bin/ld" "-demangle" "-lto_library" "/usr/local/lib/libLTO.dylib" "-no_deduplicate"
 "-dynamic" "-arch" "x86_64" "-macosx_version_min" "10.12.0" "-o" "a.out" "/var/folders/r9/
35q9g3d56_d9g0v59w9x2l9w0000gn/T/main-85975b.o" "-lSystem" "/usr/local/lib/../lib/clang/5.
0.0/lib/darwin/libclang_rt.osx.a"
```

这里的 -I 是由 clang 支持的。在 clang 里有 option 类，clang 会对这些 option 类进行专门解析，使用 DSL 语言将其转成 .tb 文件后，再使用 table-gen 工具转成 C++ 和其他代码一起进行编译。

Driver 层会解析传入的 -I Option 参数。

在 -x 后加 c 表示是对 C 语言进行编译，clang driver 是通过文件的后缀 .c 自动加上这个参数的。如果是 C++，那么仅仅使用 -x 参数来编译后缀为 .cpp 的 C++ 文件是会出错的。

```
clang -x c++ main.cpp
```

```
→ llvmdemo clang -x c++ main.cpp
Undefined symbols for architecture x86_64:
  "std::__1::__basic_string_common<true>::__throw_length_error() const", referenced from:
      std::__1::ostreambuf_iterator<char, std::__1::char_traits<char> > std::__1::__pad_an
d_output<char, std::__1::char_traits<char> >(std::__1::ostreambuf_iterator<char, std::__1:
:char_traits<char> >, char const*, char const*, char const*, std::__1::ios_base&, char) in
 main-e1e673.o
  "std::__1::locale::use_facet(std::__1::locale::id&) const", referenced from:
      std::__1::basic_ostream<char, std::__1::char_traits<char> >& std::__1::endl<char, st
d::__1::char_traits<char> >(std::__1::basic_ostream<char, std::__1::char_traits<char> >&)
 in main-e1e673.o
      std::__1::basic_ostream<char, std::__1::char_traits<char> >& std::__1::__put_charact
er_sequence<char, std::__1::char_traits<char> >(std::__1::basic_ostream<char, std::__1::ch
ar_traits<char> >&, char const*, unsigned long) in main-e1e673.o
  "std::__1::ios_base::getloc() const", referenced from:
      std::__1::basic_ostream<char, std::__1::char_traits<char> >& std::__1::endl<char, st
d::__1::char_traits<char> >(std::__1::basic_ostream<char, std::__1::char_traits<char> >&)
 in main-e1e673.o
      std::__1::basic_ostream<char, std::__1::char_traits<char> >& std::__1::__put_charact
er_sequence<char, std::__1::char_traits<char> >(std::__1::basic_ostream<char, std::__1::ch
ar_traits<char> >&, char const*, unsigned long) in main-e1e673.o
  "std::__1::basic_string<char, std::__1::char_traits<char>, std::__1::allocator<char> >::
~basic_string()", referenced from:
      std::__1::ostreambuf_iterator<char, std::__1::char_traits<char> > std::__1::__pad_an
d_output<char, std::__1::char_traits<char> >(std::__1::ostreambuf_iterator<char, std::__1:
:char_traits<char> >, char const*, char const*, char const*, std::__1::ios_base&, char) in
 main-e1e673.o
  "std::__1::basic_ostream<char, std::__1::char_traits<char> >::put(char)", referenced fro
m:
      std::__1::basic_ostream<char, std::__1::char_traits<char> >& std::__1::endl<char, st
d::__1::char_traits<char> >(std::__1::basic_ostream<char, std::__1::char_traits<char> >&)
 in main-e1e673.o
```

通过报错信息可以看出代码中存在链接错误。

因为需要链接 C++ 的标准库，所以加上参数 -lc++ 就可以了。

```
clang -x c++ -lc++ main.cpp
```

clang++ 和 clang 命令的区别就在于是否会加载 C++ 库。其实 clang++ 最终还是会调用 clang 的, 那么手动指定加载库就好了, 何必还要多个 clang++ 命令呢? 这主要是为了能够在这个命令里加载更多的库, 除标准库以外, 还有非 C++ 标准库、辅助库, 等等。因此, 针对 C++ 的程序用 clang++ 就够了。

只有加上 -cc1 才能进入 clang driver , 比如要使用 emit-obj 就需要先加上 -cc1。

如何实现 clang driver 的开关? 我们可以阅读 driver.cpp 源代码, 在 main() 函数里了解实现方式。首先进行多平台的兼容处理, 然后进行对入参的判断, 判断第一个入参是不是 -cc1。

```
if (MarkEOLs && argv.size() > 1 && StringRef(argv[1]).startswith("-cc1"))
  MarkEOLs = false;
llvm::cl::ExpandResponseFiles(Saver, Tokenizer, argv, MarkEOLs);

// 处理 -cc1 集成工具
auto FirstArg = std::find_if(argv.begin() + 1, argv.end(),
                             [](const char *A) { return A != nullptr; });
if (FirstArg != argv.end() &&
    StringRef(*FirstArg).startswith("-cc1")) {
  // 如果 -cc1 来自 response file, 移除 EOL sentinels
  if (MarkEOLs) {
    auto newEnd = std::remove(argv.begin(), argv.end(), nullptr);
    argv.resize(newEnd - argv.begin());
  }
  return ExecuteCC1Tool(argv, argv[1] + 4);
}
```

如果第一个入参是 -cc1 , 那么会调用 ExecuteCC1Tool 函数。

```
static int ExecuteCC1Tool(ArrayRef<const char *> argv, StringRef Tool) {
  void *GetExecutablePathVP = (void *)(intptr_t) GetExecutablePath;
  if (Tool == "")
    return cc1_main(argv.slice(2), argv[0], GetExecutablePathVP);
  if (Tool == "as")
    return cc1as_main(argv.slice(2), argv[0], GetExecutablePathVP);

  // 拒绝未知工具
  llvm::errs() << "error: unknown integrated tool '" << Tool << "'\n";
  return 1;
}
```

最终会执行 cc1-main 或者 cc1as_main。这两个函数分别在与 driver.cpp 同级的目录里的 cc1_main.cpp 和 cc1as_main.cpp 中。

下面我们看一看有哪些解析 Args 的方法。

- ParseAnalyzerArgs : 解析出静态分析器 option。
- ParseMigratorArgs: 解析 Migrator option。
- ParseDependencyOutputArgs: 解析依赖输出 option。
- ParseCommentArgs: 解析注释 option。
- ParseFileSystemArgs: 解析文件系统 option。
- ParseFrontendArgs: 解析前端 option。
- ParseTargetArgs : 解析目标 option。

- ParseCodeGenArgs：解析 CodeGen 的相关 option。
- ParseHeaderSearchArgs：解析 HeaderSearch 对象初始化相关的 option。
- parseSanitizerKinds：解析 Sanitizer Kinds。
- ParsePreprocessorArgs：解析预处理 option。
- ParsePreprocessorOutputArg：解析预处理输出 option。

2.7.3 Pipeline

在 Pipeline 中添加 -ccc-print-phases 选项以后就可以打印出 Pipeline 的全部功能了。

使用-ccc-print-phases 选项，在编译时会生成以 .inc 为后缀的 C++ TableGen 文件。在 Options.td 中可以看到全部选项的定义。

在 clang 的 Pipeline 中，很多实际行为都有对应的 Action，比如在 preprocessor 这个阶段提供文件的行为对应的 Action 是 InputAction，绑定机器架构的行为对应的 Action 是 BindArchAction。

执行 clang main.c -arch i386 -arch x86_64 -o main 命令后，再执行 file main 命令查看结果，就可以看到两个架构都被包含到了一个文件里，这就是 BindArchAction 的作用。

2.7.4 Action

编译前端的 Action 负责处理编译中实际要做的事情。BuildActions 和 BuildUniversalActions 是构建 Action 的入口，根据不同参数执行不同的 Action。BuildActions 和 BuildUniversalActions 的定义如下：

```
/// BuildActions - Construct the list of actions to perform for the
/// given arguments, which are only done for a single architecture.
///
/// \param C - The compilation that is being built.
/// \param Args - The input arguments.
/// \param Actions - The list to store the resulting actions onto.
void BuildActions(Compilation &C, llvm::opt::DerivedArgList &Args,
          const InputList &Inputs, ActionList &Actions) const;

/// BuildUniversalActions - Construct the list of actions to perform
/// for the given arguments, which may require a universal build.
///
/// \param C - The compilation that is being built.
/// \param TC - The default host tool chain.
void BuildUniversalActions(Compilation &C, const ToolChain &TC,
          const InputList &BAInputs) const;
```

在上面两个方法中，BuildUniversalActions 调用的还是 BuildActions。BuildActions 方法的实现如下所示。

```
void Driver::BuildActions(Compilation &C, DerivedArgList &Args,
          const InputList &Inputs, ActionList &Actions)
          const {
```

```cpp
  llvm::PrettyStackTraceString CrashInfo("Building compilation actions");

  if (!SuppressMissingInputWarning && Inputs.empty()) {
    Diag(clang::diag::err_drv_no_input_files);
    return;
  }

  Arg *FinalPhaseArg;
  phases::ID FinalPhase = getFinalPhase(Args, &FinalPhaseArg);
```

接下来看一下 getFinalPhase 方法的实现。

```cpp
// -{E,EP,P,M,MM} only run the preprocessor.
if (CCCIsCPP() || (PhaseArg = DAL.getLastArg(options::OPT_E)) ||
    (PhaseArg = DAL.getLastArg(options::OPT__SLASH_EP)) ||
    (PhaseArg = DAL.getLastArg(options::OPT_M, options::OPT_MM)) ||
    (PhaseArg = DAL.getLastArg(options::OPT__SLASH_P))) {
  FinalPhase = phases::Preprocess;

  // -{fsyntax-only,-analyze,emit-ast} only run up to the compiler.
} else if ((PhaseArg = DAL.getLastArg(options::OPT_fsyntax_only)) ||
           (PhaseArg = DAL.getLastArg(options::OPT_module_file_info)) ||
           (PhaseArg = DAL.getLastArg(options::OPT_verify_pch)) ||
           (PhaseArg = DAL.getLastArg(options::OPT_rewrite_objc)) ||
           (PhaseArg = DAL.getLastArg(options::OPT_rewrite_legacy_objc)) ||
           (PhaseArg = DAL.getLastArg(options::OPT__migrate)) ||
           (PhaseArg = DAL.getLastArg(options::OPT__analyze,
                                      options::OPT__analyze_auto)) ||
           (PhaseArg = DAL.getLastArg(options::OPT_emit_ast))) {
  FinalPhase = phases::Compile;

  // -S only runs up to the backend.
} else if ((PhaseArg = DAL.getLastArg(options::OPT_S))) {
  FinalPhase = phases::Backend;

  // -c compilation only runs up to the assembler.
} else if ((PhaseArg = DAL.getLastArg(options::OPT_c))) {
  FinalPhase = phases::Assemble;

  // Otherwise do everything.
} else
  FinalPhase = phases::Link;
```

看完这段代码我们就会发现，每次设置的 option 都会完整地走一遍流程，即从预处理、静态分析、编译后端，再到汇编。

下面是一些编译器的前端 Action。

- InitOnlyAction：只做前端初始化，对应的 option 是-init-only。
- PreprocessOnlyAction：只做预处理，不输出，对应的 option 是 -Eonly。
- PrintPreprocessedAction：做预处理，子选项包括-P、-C、-dM、-dD（具体可以查看 PreprocessorOutputOptions 这个类），对应的 option 是 -E。
- RewriteIncludesAction：预处理。

- DumpTokensAction：打印 token，对应的 option 是 -dump-tokens。
- DumpRawTokensAction：输出原始 token，包括空格，对应的 option 是 -dump-raw-tokens。
- RewriteMacrosAction：处理并扩展宏定义，对应的 option 是 -rewrite-macros。
- HTMLPrintAction：生成高亮的代码网页，对应的 option 是 -emit-html。
- DeclContextPrintAction：打印声明，对应的 option 是 -print-decl-contexts。
- ASTDeclListAction：打印语法树节点，对应的 option 是 -ast-list。
- ASTDumpAction：打印语法树详细信息，对应的 option 是 -ast-dump。
- ASTViewAction：生成语法树的 dot 文件，能够通过 Graphviz 查看图形语法树，对应的 option 是 -ast-view。
- AnalysisAction：运行静态分析引擎，对应的 option 是 -analyze。
- EmitLLVMAction：生成可读的 IR 文件，对应的 option 是 -emit-llvm。
- EmitBCAction：生成 IR Bitcode 文件，对应的 option 是 -emit-llvm-bc。
- MigrateSourceAction：代码迁移，对应的 option 是 -migrate。

2.7.5　Bind

Bind 主要与 ToolChain（工具链）交互。根据创建的 Action，Bind 会在 Action 执行过程中为 ToolChain 提供所需要的工具。比如，生成汇编时使用的工具是内嵌的工具、GNU 的工具，还是其他工具呢？这个是由 Bind 决定的，具体使用的工具由各个架构、平台、系统的 ToolChain 来决定。

使用 clang -ccc-print-bindings main.c -o main 分析 Bind，通过其结果，可以看到编译选择的工具是 clang，链接选择的工具是 darwin::Linker，但是在链接时没有使用汇编器，这是因为 Bind 起了作用，它会根据不同的平台来选择不同的工具。因为是在 Mac 平台上，所以 Bind 决定使用内置汇编器 integrated-as。那么，为何不用内置汇编器呢？答案是使用 -fno-integrated-as 这个 option。

2.7.6　Translate

Translate 是指定参数转换的工具。若在执行特定的 Action 时选择了一个工具，这个工具就必须在编译期间创建一个具体的命令，其主要工作就是将 gcc 风格的命令行选项转换成子进程所需要的选项。对于编译器或链接器，这一过程会额外转换大量的参数，但对于汇编工具，只会做少量的参数转换，仅仅需要确认调用可执行文件的路径、输入参数、输出参数即可。ArgList 类提供了大量简单的方法用于转换参数。

2.7.7　Jobs

创建 Jobs 的方法如下。

```
/// BuildJobsForAction - Construct the jobs to perform for the action
/// \p A and
/// return an InputInfo for the result of running \p A. Will only construct
/// jobs for a given (Action, ToolChain, BoundArch, DeviceKind) tuple once.
```

```
InputInfo
BuildJobsForAction(Compilation &C, const Action *A, const ToolChain *TC,
            StringRef BoundArch, bool AtTopLevel, bool MultipleArchs,
            const char *LinkingOutput,
            std::map<std::pair<const Action *, std::string>,
            InputInfo>&CachedResults, Action::OffloadKind
            TargetDeviceOffloadKind) const;
```

可以看出，Jobs 的执行需要基于前面的 Compilation、Action、ToolChain 等的运行结果。其实，Jobs 就是将前面过程获取的信息进行组合并分组给后面的 Execute 做万全准备的。

2.7.8 Execute

在 driver.cpp 的 main 函数的 ExecuteCompilation 方法中，可以看到如下代码：

```
// Set up response file names for each command, if necessary
for (auto &Job : C.getJobs())
  setUpResponseFiles(C, Job);

C.ExecuteJobs(C.getJobs(), FailingCommands);
```

准备好 Jobs 之后，就可以开始进行 Execute 了。

Execute 就是执行整个编译过程的 Jobs。执行过程的内容和耗时可以通过添加 -ftime-report 这个 option 查看。

```
→ llvmdemo clang main.c -ftime-report
===-------------------------------------------------------------------------===
                        Miscellaneous Ungrouped Timers
===-------------------------------------------------------------------------===

  ---User Time---   --System Time--   --User+System--   ---Wall Time---  --- Name ---
   0.0027 ( 77.4%)   0.0007 ( 67.9%)   0.0034 ( 75.3%)   0.0034 ( 75.1%)  Code Generation Time
   0.0008 ( 22.6%)   0.0003 ( 32.1%)   0.0011 ( 24.7%)   0.0011 ( 24.9%)  LLVM IR Generation Time
   0.0035 (100.0%)   0.0010 (100.0%)   0.0045 (100.0%)   0.0045 (100.0%)  Total

===-------------------------------------------------------------------------===
                                 DWARF Emission
===-------------------------------------------------------------------------===
  Total Execution Time: 0.0000 seconds (0.0000 wall clock)

  ---User Time---   --System Time--   --User+System--   ---Wall Time---  --- Name ---
   0.0000 ( 58.3%)   0.0000 ( 50.0%)   0.0000 ( 56.5%)   0.0000 ( 56.2%)  Debug Info Emission
   0.0000 ( 27.8%)   0.0000 ( 40.0%)   0.0000 ( 30.4%)   0.0000 ( 29.7%)  DWARF Exception Writer
   0.0000 ( 13.9%)   0.0000 ( 10.0%)   0.0000 ( 13.0%)   0.0000 ( 14.1%)  DWARF Debug Writer
   0.0000 (100.0%)   0.0000 (100.0%)   0.0000 (100.0%)   0.0000 (100.0%)  Total
===-------------------------------------------------------------------------===
```

```
              ... Pass execution timing report ...
===-------------------------------------------------------------------===
  Total Execution Time: 0.0017 seconds (0.0017 wall clock)

   ---User Time---   --System Time--  --User+System--   ---Wall Time---  --- Name ---
   0.0005 ( 38.0%)   0.0001 ( 28.2%)  0.0006 ( 35.9%)   0.0006 ( 35.9%)  Expand Atomic instructions
   0.0003 ( 22.9%)   0.0001 ( 22.3%)  0.0004 ( 22.7%)   0.0004 ( 22.7%)  X86 DAG->DAG Instruction Selection
   0.0001 ( 10.3%)   0.0000 (  9.9%)  0.0002 ( 10.3%)   0.0002 ( 10.3%)  X86 Assembly Printer
   0.0001 (  4.0%)   0.0000 (  7.8%)  0.0001 (  4.8%)   0.0001 (  4.8%)  Module Verifier
   0.0001 (  4.5%)   0.0000 (  3.5%)  0.0001 (  4.3%)   0.0001 (  4.2%)  Prologue/Epilogue Insertion & Frame Finalization
   0.0001 (  3.9%)   0.0000 (  3.5%)  0.0001 (  3.8%)   0.0001 (  3.8%)  Fast Register Allocator
   0.0000 (  1.9%)   0.0000 (  1.3%)  0.0000 (  1.7%)   0.0000 (  1.9%)  Insert stack protectors
   0.0000 (  1.6%)   0.0000 (  2.7%)  0.0000 (  1.9%)   0.0000 (  1.8%)  Inliner for always_inline functions
   0.0000 (  1.7%)   0.0000 (  2.2%)  0.0000 (  1.8%)   0.0000 (  1.7%)  Two-Address instruction pass
   0.0000 (  1.3%)   0.0000 (  2.4%)  0.0000 (  1.6%)   0.0000 (  1.5%)  CallGraph Construction
   0.0000 (  1.2%)   0.0000 (  1.6%)  0.0000 (  1.3%)   0.0000 (  1.4%)  Natural Loop Information
   0.0000 (  0.8%)   0.0000 (  1.1%)  0.0000 (  0.9%)   0.0000 (  0.9%)  Scalar Evolution Analysis
   0.0000 (  0.9%)   0.0000 (  0.0%)  0.0000 (  0.7%)   0.0000 (  0.8%)  Module Verifier
   0.0000 (  0.5%)   0.0000 (  1.1%)  0.0000 (  0.6%)   0.0000 (  0.7%)  Remove unreachable blocks from the CFG
   0.0000 (  0.7%)   0.0000 (  0.0%)  0.0000 (  0.6%)   0.0000 (  0.7%)  Module Verifier
   0.0000 (  0.4%)   0.0000 (  1.1%)  0.0000 (  0.5%)   0.0000 (  0.5%)  Post-RA pseudo instruction expansion pass
   0.0000 (  0.4%)   0.0000 (  0.3%)  0.0000 (  0.3%)   0.0000 (  0.5%)  Dominator Tree Construction
   0.0000 (  0.5%)   0.0000 (  0.0%)  0.0000 (  0.4%)   0.0000 (  0.5%)  Unnamed pass: implement Pass::getPassName()
   0.0000 (  0.4%)   0.0000 (  1.3%)  0.0000 (  0.6%)   0.0000 (  0.4%)  Machine Module Information
   0.0000 (  0.3%)   0.0000 (  0.3%)  0.0000 (  0.3%)   0.0000 (  0.4%)  Basic Alias Analysis (stateless AA impl)
   0.0000 (  0.2%)   0.0000 (  0.5%)  0.0000 (  0.3%)   0.0000 (  0.3%)  Eliminate PHI nodes for register allocation
   0.0000 (  0.2%)   0.0000 (  0.8%)  0.0000 (  0.3%)   0.0000 (  0.3%)  Bundle Machine CFG Edges
   0.0000 (  0.1%)   0.0000 (  0.8%)  0.0000 (  0.3%)   0.0000 (  0.3%)  Function Alias Analysis Results
   0.0000 (  0.3%)   0.0000 (  0.0%)  0.0000 (  0.2%)   0.0000 (  0.3%)  Dominator Tree Construction
   0.0000 (  0.1%)   0.0000 (  0.3%)  0.0000 (  0.2%)   0.0000 (  0.2%)  Inserts calls to mcount-like functions
   0.0000 (  0.1%)   0.0000 (  0.5%)  0.0000 (  0.2%)   0.0000 (  0.2%)  Safe Stack instrumentation pass
```

2.8 clang attribute

clang attribute 是 clang 提供的能够让开发者在编译过程中参与源代码控制的方法，语法格式为 attribute（xx）。下面介绍 clang attribute 的用法。

format：格式化字符串。

可以查看 NSLog 的用法。

```
FOUNDATION_EXPORT void NSLog(NSString *format, …)
NS_FORMAT_FUNCTION(1,2) NS_NO_TAIL_CALL;

// Marks APIs which format strings by taking a format string and optional
// varargs as arguments
#if !defined(NS_FORMAT_FUNCTION)
    #if (__GNUC__ *10+__GNUC_MINOR__ >= 42) && (TARGET_OS_MAC ||
        TARGET_OS_EMBEDDED)
     #define NS_FORMAT_FUNCTION(F,A) __attribute__((format(__NSString__,
        F, A)))
    #else
     #define NS_FORMAT_FUNCTION(F,A)
    #endif
#endif
#deprecated 版本弃用提示
```

在编译过程中提示开发者某方法或某属性已经被弃用。

```
- (void)preMethod:( NSString *)string __attribute__((deprecated("preMethod 已经
被弃用，请使用 newMethod")));
- (void)deprecatedMethod DEPRECATED_ATTRIBUTE; //也可以直接使用
DEPRECATED_ATTRIBUTE 这个系统定义的宏
```

availability：指明使用版本范围。

os 指系统的版本，m 指引入的版本，n 指过时的版本，o 指完全不用的版本，message 指可以写入的描述信息。

```
- (void)method __attribute__((availability(ios,introduced=3_0,
deprecated=6_0,obsoleted=7_0,message="iOS3 到 iOS7 版本可用，iOS7 不能用")));
```

unavailable：方法不可用提示。

当编写代码时，有时系统会提示使用了不可用的方法，会导致编译失败。

unused：静态函数即使不使用也不报警。

默认情况下，在代码中定义了一个静态函数，若不使用，在编译时会显示警告。如果使用 unused 属性修饰定义的静态函数，那么即使不使用定义的静态函数，在编译时也不会报警。

warn_unused_result：调用者必须使用函数返回值。

当调用 warn_unused_result 属性修饰函数时，调用者必须使用返回值，否则编译器会出现警告提示。在 Swift 3 之后的版本中，所有函数默认支持这个 clang attribute。

cleanup：作用域结束时自动执行一个指定方法。

作用域结束，包括 right brace、return、goto、break、exception 等情况。结束动作是先于使用对象的 dealloc 调用的。

Reactive Cocoa 中比较好的使用范例是@onExit 宏，定义如下：

```
#define onExit \
    rac_keywordify \
    __strong rac_cleanupBlock_t metamacro_concat(rac_exitBlock_, __LINE__) 
__attribute__((cleanup(rac_executeCleanupBlock), unused)) = ^

static inline void rac_executeCleanupBlock (__strong
                                            rac_cleanupBlock_t *block) {
    (*block)();
}
```

这样就可以很方便地把需要成对出现的代码写在一起了，并且可以在 Reactive Cocoa 中看到具体应用。

```
if (property != NULL) {
        rac_propertyAttributes *attributes = rac_copyPropertyAttributes
                                            (property);
        if (attributes != NULL) {
            @onExit {
                free(attributes);
            };

            BOOL isObject = attributes->objectClass != nil || strstr
                            (attributes->type, @encode(id)) ==
                            attributes->type;
            BOOL isProtocol = attributes->objectClass ==
                              NSClassFromString(@"Protocol");
            BOOL isBlock = strcmp(attributes->type,
                           @encode(void(^)())) == 0;
```

```
            BOOL isWeak = attributes->weak;

            shouldAddDeallocObserver = isObject && isWeak
                                     && !isBlock && !isProtocol;
    }
}
```

可以看出，attributes 的设置和释放都在一起，使得代码的可读性得到了提升。

overloadable：方法重载。

在 C 语言的函数上实现函数重载，即具有相同函数名的函数可以有多个重载版本，能够实现一个函数接受不同参数类型的效果。

```
__attribute__((overloadable)) void printArgument(int number){
  NSLog(@"Add Int %i", number);
}

__attribute__((overloadable)) void printArgument(NSString *number){
  NSLog(@"Add NSString %@", number);
}

__attribute__((overloadable)) void printArgument(NSNumber *number){
  NSLog(@"Add NSNumber %@", number);
}
```

objc_designated_initializer：指定内部实现的初始化方法。

- 如果是用 objc_designated_initializer 初始化的方法，那么必须调用覆盖实现 super 的 objc_designated_initializer 方法。
- 如果不是用 objc_designated_initializer 初始化的方法，但是该类拥有 objc_designated_initializer 初始化的方法，那么必须调用该类的 objc_designated_initializer 方法或者非 objc_designated_initializer 方法，而不能调用 super 的任何初始化方法。

objc_subclassing_restricted：指定不能有子类。

相当于 Java 的 final 关键字，如果有子类，继承时就会出错。

objc_requires_super：子类继承必须调用 super。

声明后子类在继承这个方法时必须调用 super，否则会出现编译警告。在 super 里定义一些必要的执行方法，并通过这个属性提醒使用者定义这个方法是必要的。

const：重复调用相同数值参数时的优化处理。

这个属性用于参数为数值类型的函数，多次调用相同的数值型参数，返回结果是相同的，所以只需在第一次调用时进行运算，以后再调用时返回第一次调用的运算结果，这是编译器的一种优化处理方式。

constructor 和 destructor。

程序在执行 main 函数之前会执行 constructor，执行 main 函数之后会执行 destructor。

+load 方法会在 constructor 之前执行，因为动态链接器加载 Mach-O 文件时会先加载每个类，需要调用 +load 方法，然后才会调用所有的 constructor 方法。

通过这个特性，可以实现一些有趣的功能，比如某个类完成 load 之后，是可以在 constructor 中对想替换的类进行替换的，而不用把相关代码加在特定类的 +load 方法里。

2.9　clang 警告处理

我们看一看以下案例。

```
#pragma clang diagnostic push
#pragma clang diagnostic ignored "-Wdeprecated-declarations"
     sizeLabel = [self sizeWithFont:font constrainedToSize:size
                  lineBreakMode:NSLineBreakByWordWrapping];
#pragma clang diagnostic pop
```

如果没有 #pragma clang diagnostic push、#pragma clang diagnostic ignored "-Wdeprecated-declarations"、#pragma clang diagnostic pop 定义，编译器会发出 sizeWithFont 方法被废弃的警告。在 Label 中加上 sizeWithFont:constrainedToSize:lineBreakMode: 方法是为了兼容老系统，加上 ignored "-Wdeprecated-declarations" 的作用是忽略这个警告。通过 clang diagnostic push/pop，可以灵活地控制代码块的编译选项。

2.10　通过 LibTooling 控制语法树

因为 LibTooling 能够完全控制语法树，所以可以做的事情就非常多了。
- 改变 clang 生成代码的方式。
- 增加更强的类型检查。
- 按照定义进行代码的检查分析。
- 对源代码做任意类型分析，甚至重写程序。
- 给 clang 添加一些自定义的分析，创建自己的重构器。
- 基于现有代码做出大量的修改。
- 基于工程生成相关图形或文档。
- 检查命名是否规范。LibTooling 还能够进行语言的转换，比如把 Objective-C 转成 JavaScript 或者 Swift 。

开发者可以按照官方文档的说明去构造 LLVM、clang 和其他工具。

按照说明编译完成后，进入 LLVM 的目录 ~/llvm/tools/clang/tools/创建自己的 clang 工具。

下面是检查 Target 对象中是否有对应的 Action 方法存在的例子。

```
@interface Observer
+ (instancetype)observerWithTarget:(id)target action:(SEL)selector;
@end
//查找消息表达式，observer 作为接受者，observerWithTarget:action: 作为
//selector，检查 Target 中是否存在相应的方法。
```

```cpp
    virtual bool VisitObjCMessageExpr(ObjCMessageExpr *E) {
      if (E->getReceiverKind() == ObjCMessageExpr::Class) {
        QualType ReceiverType = E->getClassReceiver();
        Selector Sel = E->getSelector();
        string TypeName = ReceiverType.getAsString();
        string SelName = Sel.getAsString();
        if (TypeName == "Observer" && SelName == "observerWithTarget:action:") {
          Expr *Receiver = E->getArg(0)->IgnoreParenCasts();
          ObjCSelectorExpr* SelExpr =
cast<ObjCSelectorExpr>(E->getArg(1)->IgnoreParenCasts());
          Selector Sel = SelExpr->getSelector();
          if (const ObjCObjectPointerType *OT = Receiver->getType()->getAs
                                    <ObjCObjectPointerType>()) {
            ObjCInterfaceDecl *decl = OT->getInterfaceDecl();
            if (! decl->lookupInstanceMethod(Sel)) {
              errs() << "Warning: class " << TypeName <<
              " does not implement selector " << Sel.getAsString() << "\n";
              SourceLocation Loc = E->getExprLoc();
              PresumedLoc PLoc = astContext->getSourceManager()
                                .getPresumedLoc(Loc);
              errs() << "in " << PLoc.getFilename() << " <" << PLoc.getLine()
              << ":" << PLoc.getColumn() << ">\n";
            }
          }
        }
      }
      return true;
    }
```

2.11 clang 插件

通过编写插件，比如上面提到的 LibTooling 的 clang 工具，可以将相应插件动态加载到编译器中，从而对编译过程进行控制。可以在 LLVM 的目录下查看一些范例（llvm/tools/clang/tools）。

滴滴出行的王康在做包大小瘦身时，曾实现了一个自定义的 clang 插件，自定义插件的具体实现过程可以查看他的文章《基于 clang 插件的一种 iOS 包大小瘦身方案》。

2.12 LLVM Backend

clang 基于源代码生成 IR 之后，将 IR 转成目标机器码的过程就是 LLVM Backend。下图列出了整个 LLVM 的编译过程，其中右下角黑框内描述的过程就是 LLVM Backend。

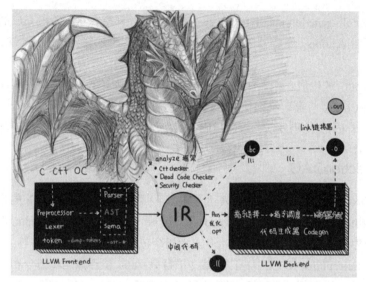

接下来,看一下整个 LLVM Backend 的流程图,如下图所示,包括 SelectionDAG(包含 Lower、DAG Combine1、Legalize、DAG Combine2、ISel)、Register Allocation(下图中的 Instruction Scheduling/Pre-RA、Register Allocation、Instruction Scheduling/Post-RA)、Code Emission(下图中的 Object File Assembler BinaryCode、MCStream)。后面会对每个过程进行详细说明。

2.12.1 CodeGen 阶段

CodeGen 是完整的 LLVM Backend 阶段的描述,整个过程分为以下几个阶段。
- Instruction Selection:指令选择,将 IR 转化成由目标平台指令组成的 DAG(Directed Acyclic Graph,定向非循环图)。选择能完成指定操作且执行时间最短的指令。
- Scheduling and Formation:调度与排序,读取 DAG,将 DAG 的指令排成 MachineInstr 队列。根据指令间的依赖进行指令重排,使其更好地利用 CPU 的功能单元。
- SSA 优化:由多个基于 SSA(Static Single Assignment,静态单赋值)的 Pass 组成。

比如 modulo-scheduling 和 peephole optimization 都是在这个阶段完成的。
- Register allocation：寄存器分配，将 Virtual Register 映射到 Physical Register 或者内存地址上。
- Prolog/Epilo 的生成：函数体指令生成后，就可以确定函数所需要的堆栈大小了。
- Machine Code：机器码晚期优化，这是最后一次进行优化的机会。
- Code Emission：代码发射，输出代码，可以选择输出汇编程序或二进制机器码。

2.12.2 SelectionDAG

SelectionDAG 是代码的一种抽象表达方式。下面列出 SelectionDAG 的指令选择过程。
- 构建最初的 DAG：把 IR 里的 add 指令转成 SelectionDAG 的 add 节点。
- 优化构建好的 DAG：识别出元指令。元指令包括 rotate、div/rem 等。
- 合法化 SelectionDAG 类型：比如某些平台只有 64 位浮点运算指令和 32 位整数运算指令，那么就需要把所有 f32 都提升到 f64，把 i1/i8/i16 都提升到 i32。同时还要把 i64 拆分成两个 i32 来存储。此过程还包括操作符的合法化，比如 SDIV 在 x86 上回转成 SDIVREM。整个过程的结果可以通过 llc -view-dag-combine2-dags sum.ll 看到。
- 指令选择：将平台无关的 DAG 通过 TableGen 读入 .tb 文件，并且生成对应的模式匹配代码，从而转成平台相关的 DAG。
- SelectionDAG 重排指令队列：因为 CPU 是无法执行 DAG 的，所以从 DAG 中提取指令时需要依据一定规则，比如最小寄存器压力、隐藏指令延迟按照 DAG → linear list（SSA form）→ MachineInstr → MC Layer API MCInst MCStreamr → MCCodeEmitter → Binary Instr 顺序排列。

下图是执行完 llc -view-isel-dags sum.ll 后的 DAG 图，表示的是 DAG 在 -view-isel-dags 状态的情况。

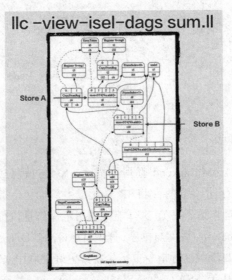

查看 DAG 不同状态的说明如下：

- -view-dag-combine1-dags：显示没有被优化的 DAG。
- -view-legalize-dags：显示合法化之前的 DAG。
- -view-dag-cmobine2-dags：显示第二次优化前的 DAG。
- -view-isel-dags：显示指令选择前的 DAG。
- -view-sched-dags：显示在 ISel 阶段后、Scheduler 阶段前的 DAG。
- -view-sunit-dags：显示 Scheduler 的依赖图。

SDNode

DAG 的节点都是由 SDNode 构成的，主要是作为 DAG 值的操作符，描述 DAG 所代表的操作数。SDNode 的定义代码存储在 LLVM 源代码的 SelectDAGNodes.h 文件里，还有些定义代码存储在 TargetSelectionDAG.td 里。每个 SelectionDAG 节点类型都有一个对应的 SDNode 定义。

```
class SDNode<string opcode, SDTypeProfile typeprof, list<SDNodeProperty>
 props = [], string sdclass ="SDNode"> :SDPatternOperator {
    stringOpcode = opcode;
    string SDClass= sdclass;
    list<SDNodeProperty> Properties = props;
    SDTypeProfileTypeProfile = typeprof; //类型
}
//类型要求
class SDTypeProfile<intnumresults, int numoperands,
 list<SDTypeConstraint>constraints> {
    int NumResults= numresults;   //多少个结果
    int NumOperands= numoperands; //多少个操作数
    list<SDTypeConstraint> Constraints = constraints; //类型的约束
}
//描述对操作数类型的约束
class SDTypeConstraint<intopnum> {
    int OperandNum= opnum; //指明该约束适用第几个操作数
}
```

目标机器可以根据需求定制约束，来描述特有的指令。

SDNodeProperty 是 SDNode 的属性，用来描述 SDNode 操作的特征。

PatFrag

PatFrag 指可复用的结构。为了支持高级语言的特性，TD 语言（目标描述文件里使用的语言）可以通过 PatFrag 达到对高级语言特性的支持。这样就可以支持数据与结构的复用了。

```
class PatFrag<dag ops, dag frag, code pred = [{}],
  SDNodeXForm xform =NOOP_SDNodeXForm> : SDPatternOperator {
    dag Operands= ops;   //操作数
    dag Fragment= frag;  //所指定的片段
    code PredicateCode = pred; //表示嵌入生成的 ISel 代码,满足条件采用 PatFrag
    code ImmediateCode = [{}];
    SDNodeXForm OperandTransform = xform;
}
```

Pattern

Pattern 指匹配指令，主要用于解决 DAG 模式的复杂操作问题，LLVM 会使用贪婪匹配算法自动完成 Pattern 指令对的匹配。Pattern 的定义在 Target.td 里。

```
class Pattern<dag patternToMatch, list<dag>resultInstrs> {
    dagPatternToMatch = patternToMatch;
    list<dag>ResultInstrs = resultInstrs;
    list<Predicate> Predicates = [];
    int AddedComplexity = 0;
}
```

Predicate

在 Pattern 和 Instruction 的定义里都有 Predicate。满足 Predicate 的条件才能够继续执行。Predicate 的定义也在 Target.td 里。

```
class Predicate<string cond> {
    string CondString = cond;
    //汇编匹配器的 Predicate
    bit AssemblerMatcherPredicate = 0;
    //被测试的子 Target 的名称用作汇编匹配器的替代条件字符串
    string AssemblerCondString = "";
    //用户级别的 name 主要用在诊断汇编匹配器里缺少的功能
    string PredicateName = "";
}
```

这个 Predicate 实际上就是一个容器，用于装嵌代码，把代码插入合适的位置来对某些指令进行筛选过滤。

Itinerary 和 SchedRW

Itinerary 和 SchedRW 是在 Instruction 里定义的，用来描述指令调度的信息。目标机器平台会从 InstrItinClass 派生对应指令的定义。比如 x86 架构，它的指令多而且复杂，所以定义的 InstrItinClass 派生定义数量很多都在 x86Schedule.td 里，每条指令对应一个 InstrItinClass 定义。例如，除法的 InstrItinClass 定义如下：

```
def IIC_DIV8_MEM   : InstrItinClass;
def IIC_DIV8_REG   : InstrItinClass;
def IIC_DIV16      : InstrItinClass;
def IIC_DIV32      : InstrItinClass;
defIIC_DIV64       : InstrItinClass;
```

执行步骤由 InstrStage 进行描述：

```
class InstrStage<int cycles, list<FuncUnit> units, int timeinc = -1,
  ReservationKind kind =Required> {
    int Cycles          = cycles;      //完成这个步骤需要的周期数
    list<FuncUnit> Units = units;      //用于完成该步骤功能单元的选择
    int TimeInc         = timeinc;     //从此步骤到下个步骤需要多少个周期
    int Kind            = kind.Value;
}
```

通过 InstrItinData 将 InstrItinClass 和 stages 绑在一起，使得指令能按顺序执行。

```
class InstrItinData<InstrItinClass Class,list<InstrStage> stages,
list<int>operandcycles = [], list<Bypass> bypasses= [], int uops = 1> {
    InstrItinClass TheClass = Class;
    //表示指令解码后 mirco operation 的数量，0 表示数量不定
    int NumMicroOps = uops;
    list<InstrStage> Stages = stages;
    list<int> OperandCycles =operandcycles; //可选周期数
    //绕过寄存器，将写操作指令的结果直接交给后面的读操作
    list<Bypass> Bypasses = bypasses;
}
```

TableGen

在 llvm/lib/Target 目录下有 CPU 架构目录。以 x86 为例，目录名就是 x86，以下是 x86 目录包含的文件。

- x86.td：架构描述。
- x86CallingConv.td：架构调用规范。
- x86InstrInfo.td：基本指令集。
- x86InstrMMX.td：MMX 指令集。
- x86InstrMPX.td：MPX 指令集。
- x86InstrSGX.td：SGX 指令集。
- x86InstrSSE.td：SSE 指令集。
- x86InstrSVM.td：AMD SVM 指令集。
- x86InstrTSX.td：TSX 指令集。
- x86InstrVMX.td：VMX 指令集。
- x86InstrSystem.td：特权指令集。
- x86InstrXOP.td：对扩展操作的描述。
- x86InstrFMA.td：对融合乘加指令的描述。
- x86InstrFormat.td：对格式定义的描述。
- x86InstrFPStack.td：对浮点单元指令集的描述。
- x86InstrExtension.td：对零及符号扩展的描述。
- x86InstrFragmentsSIMD.td：描述 SIMD 所使用的模式片段。
- x86InstrShiftRotate.td：对 shift 和 rotate 指令的描述。
- x86Instr3DNow.td：对 3DNow! 指令集的描述。
- x86InstrArithmetic.td：对算术指令的描述。
- x86InstrAVX512.td：对 AVX512 指令集的描述。
- x86InstrCMovSetCC.td：对移动条件及设置移动条件指令的描述。
- x86InstrCompiler.td：编译器使用的各种伪指令，还有在指令选择过程中使用的 Pat 模式。
- x86InstrControl.td：对 x86 架构的 jump、return、call 指令的描述。

- x86RegisterInfo.td：对寄存器的描述。
- x86SchedHaswell.td：对 Haswell 机器模型的描述。
- x86SchedSandyBridge.td：对 Sandy Bridge 机器模型的描述。
- x86Schedule.td：对指令调度的一般描述。
- x86ScheduleAtom.td：用于 Intel Atom 处理器的指令调度。
- x86SchedSandyBridge.td：用于 Sandy Bridge 机器模型的指令调度。
- x86SchedHaswell.td：用于 Haswell 机器模型的指令调度。
- x86ScheduleSLM.td：用于 Intel Silvermont 机器模型的指令调度。
- x 86ScheduleBtVer2.td：用于 AMD btver2 (Jaguar) 机器模型的指令调度。

与平台无关的通用描述在 llvm/include/llvm/target/ 下。

- Target.td：每个机器都要实现的与平台无关的接口。
- TargetItinerary.td：与平台无关的 Instruction Itinerary 调度接口。
- TargetSchedule.td：与平台无关的基于 TableGen 的调度接口。
- TargetSelectionDAG.td：与平台无关的 SelectionDAG 调度接口。
- TargetCallingConv.td：与平台无关的 CallingConv 接口。

llvm/include/llvm/CodeGen 目录包含 ValueTypes.td，用来描述具有通用性的寄存器和操作数的类型。在 llvm/include/llvm/IR 目录中，包含描述与平台无关的固有函数文件 Intrinsics.td，以及与平台相关的文件，比如 Intrinsicsx86.td。

TableGen 包含以下类型。

- Dag：编译生成的树状结构为 DAG 结构，即递归结构。包括 DagArg 和 DagArgList，分别对应 DagArgList ::= DagArg ("," DagArg)*、DagArg ::= Value [":" TokVarName] | TokVarName 语法。比如 (set VR128:$dst, (v2i64 (scalar_to_vector (i64 (bitconvert (x86mmx VR64:$src)))))) 这个 Dag 值有多层嵌套，表达的意思是将 64 位标量的源操作数保存在 MMX 寄存器中，先转成 64 位有符号整数，再转成 2xi64 向量，保存到 128 位寄存器中。Dag 包含操作符号，比如 def、out、in、set 等，还包含 SDNode，比如 scalar_to_vector 和 bitconvert，或者 ValueType 的派生定义描述值类型，比如 VR128、i64、x86mmx 等。
- List：代表队列。
- String：C++ 字符串常量。
- Bit：代表字节。
- int：表示 64 位整数。
- Bits：代表若干字节，比如 bits<64>。

2.12.3　Register Allocation

下面从几个方面介绍 LLVM 的 Register Allocation。

寄存器

寄存器的定义代码在 TargetRegisterInfo.td 文件里，其基类定义如下：

```
class Register<string n, list<string> altNames =[]> {
    string Namespace = "";
    string AsmName = n;
    list<string> AltNames = altNames;
    //别名寄存器
    list<Register> Aliases = [];
    //属于寄存器的子寄存器
    list<Register> SubRegs = [];
    //子寄存器的索引编号
    list<SubRegIndex> SubRegIndices = [];
    //可选名寄存器的索引
    list<RegAltNameIndex> RegAltNameIndices= [];
    //gcc/gdb 定义的号码
    list<int> DwarfNumbers = [];
    //寄存器中的分配器会通过这个值尽量减少一个寄存器的指令数量
    int CostPerUse = 0;
    //决定寄存器的值是否由子寄存器的值来决定
    bit CoveredBySubRegs = 0;
    //特定硬件的编码
    bits<16> HWEncoding = 0;
}
```

根据目标机器可以派生出对应的 .td 文件，比如 x86 架构可以派生出 x86RegisterInfo.td。

```
class x86Reg<string n, bits<16> Enc, list<Register>subregs = []> :
  Register<n> {
  let Namespace= "x86";
  letHWEncoding = Enc;
  let SubRegs =subregs;
}
#RegisterClass
```

为了描述寄存器的用途，我们将相同用途的寄存器归入同一个 RegisterClass 中。

```
class RegisterClass<string namespace, list<ValueType>regTypes, int
  alignment, dagregList, RegAltNameIndex idx = NoRegAltName> : DAGOperand {
    string Namespace = namespace;

    //寄存器的值类型。寄存器里的寄存器必须有相同的值类型
    list<ValueType> RegTypes = regTypes;

    //指定寄存器溢出大小
    int Size = 0;

    //当寄存器进行存储或者读取时指定排序
    int Alignment = alignment;

    //指定在两个寄存器之间拷贝时的消耗，默认值是 1，意味着使用一个指令执行拷贝
    //如果是负数，则意味着拷贝消耗巨大或者不可拷贝
    int CopyCost = 1;
```

```
        //说明这个 class 里有哪些寄存器。如果没有指定 allocation_order_* 方法，
        //则 class 同时定义寄存器分配器的分配顺序
        dagMemberList = regList;

        //寄存器备用名用于打印操作这个寄存器 class 时
        //每个寄存器都需要在一个给定的索引里有一个有效的备用名
        RegAltNameIndex altNameIndex = idx;

        //指定寄存器 class 是否能用在虚拟寄存器和寄存器分配里
        //有些寄存器 class 只限于模型指令操作，这样就需要设置为 0
        bit isAllocatable = 1;

        //列出可选的分配命令，默认的命令是 memberlist
        //寄存器中的分配器会自动移除保留的寄存器，同时保证保留的寄存器一直是可使用的
        list<dag>AltOrders = [];

        //这个函数的作用是指定机器函数的顺序
        code AltOrderSelect = [{}];

        //寄存器中的分配器使用贪婪启发式算法指定分配优先级，值高表示优先级高
        //取值范围为[0,63]
        int AllocationPriority = 0;
}
```

寄存器在 LLVM 中的表达

在 LLVM 中，物理寄存器的编号取值范围为 1 ~ 1023。我们可以在 GenRegisterNames.inc 里找到物理寄存器的编号，比如 lib/Target/x86/x86GenRegisterInfo.inc。

从虚拟寄存器到物理寄存器的映射

在 LLVM 中，从虚拟寄存器到物理寄存器的直接映射使用的是 TargetRegisterInfo 和 MachineOperand 的 API，间接映射的 API 使用的是 VirtRegMap，通过正确插入读写指令来实现内存调度。

LLVM 自带的寄存器分配算法

LLVM 的寄存器分配指令为：

```
llc -regalloc=Greedy add.bc -o ln.s
```

下面列出分配指令的选项参数：
- Fast：编译的默认选项，尽可能保存寄存器。
- Basic：增量分配。
- Greedy：LLVM 的默认寄存器分配算法，对 Basic 算法的变量生存期进行分裂及高度优化。
- PBQP：将寄存器分配描述成分区布尔二次规划。

2.12.4 Code Emission

Code Emission 全过程为：ASMPrinter → MCInst → MCStreamer → Assembler →

MCCodeEmitter → BinaryIntr，如下图所示。

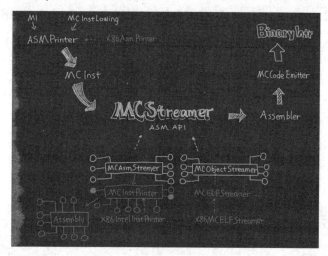

2.13 LLVM 优化

LLVM 会在以下几个阶段做优化。
- 编译阶段：编译前端将源代码编译成 LLVM IR。
- 链接阶段：跨文件分析优化。
- 运行阶段：收集运行时的分析信息。
- 闲时阶段：获得运行时的分析信息，然后生成更高效的代码。

LLVM 的优化和转换是由多个 Pass 完成的，每个 Pass 完成特定的优化工作。Pass 之间有关联，即在运行某些 Pass 之前需要先运行其他 Pass，相应地都有实现的接口。还可以对具有特定功能的 Pass 进行分组，允许将一个 Pass 注册到多个分组中。LLVM 允许开发者开发 Pass 或关闭默认的 Pass。开发或调试自定义 Pass，与系统自带的 Pass 是分开的，不会造成影响。

下面是 Pass 的分类。
- ImmutablePass：仅用于处理固定信息，并且提供当前编译器的信息。
- ModulePass：最常用的基类，能够让 Pass 将整个程序作为一个单位运行，并且可以引用、添加和删除函数。
- CallGraphSCCPass：自上而下遍历程序，分析调用关系图。
- FunctionPass：修改函数的行为，但是不能修改外部功能，包括添加或删除全局变量等。
- LoopPass：以循环嵌套顺序处理循环，最外层最后处理。
- RegionPass：和 LoopPass 类似，不同的是 RegionPass 是在函数里的单个条目退出 region 上执行的。
- BasicBlockPass：和 FunctionPass 类似，BasicBlockPass 会限制变量修改范围。
- MachineFunctionPass：是 LLVM 代码生成器的一部分，用于描述每个程序里的函数执行后在机器上的表现。

LLVM 在链接时做以下优化。

- DSA（Data Structure Analysis，数据结构分析）可以构造出整个内存对象的连接关系图，将树、链表等结构体分配在连续的内存池里，减少维护的内存块。
- APA（Automatic Pool Allocation，自动分配池）能够将分配的链接形式的结构体替换成自定义池，分配函数会把内存分配到连续的内存池中。
- 函数在进行 inline 时会删除死全局变量、死实参、死类型。
- 对常量进行传播以消除列表边界的检查。
- 对简单结构体域进行重排。
- 添加自动分配池。

下图展现了结构体域重排后的效果。

2.14 Swift 编译

Swift 编译和 clang 一样都是编译器的前端，在此阶段代码会被解析成语法树。接下来，Swift 会比 clang 多一个代码生成阶段，即通过 SILGen 生成 SIL，方便针对 Swift 做特定的优化。SIL 会被传递给 IR 生成阶段，然后生成 LLVM IR，最后由 LLVM 解决余下的问题。看到这里大家肯定会好奇，Swift 是如何与 C 语言和 Objective-C 交互的呢？比如系统底层的模块是如何交互的。这里就要说下 Swift 的模块映射（Module Map）了，它调用 clang 的模块，将其传入 clang importer 中，通过生成语法树结构进行分析，使 Swift 能够和 C 语言、Objective-C 语言进行交互。

下面通过一个例子详细了解一下 Swift 编译。首先创建一个 toy.swift：

```
print("hi!")
```

生成程序：

```
swiftc toy.swift
./toy
```

生成检查语法树：

```
swiftc -dump-ast toy.swift
```

还原之前的函数名:

```
swiftc -emit-silgen toy.swift | xcrun swift-demangle
```

LLVM IR 和汇编的生成:
```
swiftc -emit-ir toy.swift
swiftc -emit-assembly toy.swift
```

生成可执行的脚本:

```
xcrun -sdk macosx swiftc toy.swift -o toy
```

2.15 编译后生成的二进制内容 Link Map File

在 Build Settings 里设置 Write Link Map File 选项为 Yes 后,每次编译都会在指定目录生成一个文件。文件内容包含 Object file、Section 和 Symbol。下面分别讲解相关内容。

Object file

Object file 这部分内容是 .m 文件编译后的 .o 文件和需要链接的 .a 文件,前面是文件编号,后面是文件路径。

Section

Section 描述的是每个 Section 在可执行文件中的位置和大小。每个 Section 的 Segment 类型分为 __TEXT 代码段和 __DATA 数据段两种。

Symbol

Symbol 对 Section 进行了再划分,并且描述了所有的 method、ivar、string,以及它们对应的 address、size、file number 信息。

2.16 编译后生成的 dSYM 文件

在每次编译后,LLVM 都会生成一个 dSYM 文件。程序在执行过程中通过地址来调用方法函数,而 dSYM 文件里存储了函数地址映射,这样调用栈里的地址可以通过 dSYM 映射表获得具体函数的位置。dSYM 一般是用来做 .crash 文件的符号化的。.crash 是应用程序在崩溃时生成包含方法调用栈的文件。

通过 Xcode 进行符号化,将 .crash 文件、.dSYM 和 .app 文件放到同一个目录下,打开 Xcode 的 Window 菜单下,选择 Organizer 选项,在打开的窗口中单击 Device 图标,然后在左栏中单击的 Device Log 选项。单击 import 图标将 .crash 文件导入,就可以看到 crash 的详细日志了。

还可以通过命令行工具 symbolicatecrash 来手动符号化 Crash Log。同样先将 .crash 文件、.dSYM 和 .app 文件放到同一个目录下,然后输入下面的命令:

```
export DEVELOPER_DIR=/Applications/Xcode.app/Contents/Developer
```

```
symbolicatecrash appName.crash appName.app > appName.log
```

2.17 Mach-O 文件

Mach-O 文件用于记录编译后的可执行文件、对象代码、共享库、动态加载代码和内存转储的文件格式。不同于 xml 这样的文件，Mach-O 文件只是二进制字节流，里面有不同的包含元信息的数据块，比如字节顺序、CPU 类型、块大小等。Mach-O 文件的内容是不可以修改的，因为在 .app 目录中有一个 _CodeSignature 目录，里面包含了程序代码的签名，这个签名的作用就是保证.app 里的文件，包括资源文件、Mach-O 文件都不能被更改。

Mach-O 文件包含以下三个组成部分。

- Mach-O Header：包含字节顺序、魔数、CPU 类型、加载指令的数量等。
- Load Command：包括区域的位置、符号表、动态符号表等。每个加载指令包含一个元信息，比如指令类型、名称，以及在二进制中的位置等。
- Data：内容最多的部分，包含了代码、数据，比如符号表、动态符号表的数据都存储在 Data 里。

要分析 Mach-O 文件，需要先安装 tweak。对于已有 Root 权限的机器，可以通过 cydia 安装 tweak；对于，无 Root 权限的机器，直接打包成 ipa 安装包。对于有 Root 权限的机器，会安装一个 MobileSubstrate 动态库，然后使用 theos 开发工具；对于无 Root 权限的机器，直接把这个库打包进 ipa 中，或者直接修改汇编代码。

MobileSubstrate 提供了以下三个模块。

- MobileHooker：利用 method swizzling 技术定义一些宏和函数来替换系统或者目标函数。
- MobileLoader：用于在程序启动时将制定的第三方库加载到主程序运行的环境中。怎么加载呢？还记得先前介绍的 clang attribute 的 attribute((constructor)) 吗？它会在 main 执行之前执行，所以把 Hook 放在 attribute((constructor)) 里就可以了。
- Safe Mode：类似安全模式，会禁用改动。

先前提到 Mach-O 的结构有 Header、Load Command 和 Data。MobileLoader 会通过修改二进制的 LoadCommand 把 MobileLoader 自身的文件注入主程序中，然后再把我们写的第三方库注入主程序。这样，我们自己的程序就会被放入 Load Command 段了。

当然，对于自己的程序，我们是知道要替换哪些函数的，如果是分析别人的程序，这个时候就需要思考"想替换哪个方法"。对于网络相关的分析，可以用常用的抓包工具，比如 Charles、WireShark 等；对于静态的分析，可以通过砸壳、反汇编、导出头文件来分析 App 的架构，对应的常用工具有 dumpdecrypted、hopper disassembler 和 class_dump。运行时的分析工具有控制台工具 cycript、远程断点调试工具 lldb+debugserver、跟踪函数调用工具 logify。

2.18 如何利用 Mach-O

利用 Mach-O 的结构和信息能够做很多的事情，下面详细介绍这些功能。

2.18.1 打印堆栈信息，保存现场

打印堆栈信息的整体思路是获取线程的信息得到线程的状态，从而得到线程里所有栈的指针，再根据指针在符号表里的位置找到对应的描述即符号化解析。至此就能够展示出可读的堆栈信息了。

获取线程的信息

通过 task_threads 获取所有线程。

```
thread_act_array_t threads; //int 组成的数组，比如 thread[1] = 5635
//mach_msg_type_number_t 是 int 类型
mach_msg_type_number_t thread_count = 0;
const task_t this_task = mach_task_self(); //int
//根据当前 task 获取所有线程
kern_return_t kr = task_threads(this_task, &threads, &thread_count);
```

遍历时通过 thread_info 获取各个线程的详细信息。

```
SMThreadInfoStruct threadInfoSt = {0};
thread_info_data_t threadInfo;
thread_basic_info_t threadBasicInfo;
mach_msg_type_number_t threadInfoCount = THREAD_INFO_MAX;

if (thread_info((thread_act_t)thread, THREAD_BASIC_INFO,
(thread_info_t)threadInfo, &threadInfoCount) == KERN_SUCCESS) {
    threadBasicInfo = (thread_basic_info_t)threadInfo;
    if (!(threadBasicInfo->flags & TH_FLAGS_IDLE)) {
        threadInfoSt.cpuUsage = threadBasicInfo->cpu_usage / 10;
        threadInfoSt.userTime = threadBasicInfo->system_time.microseconds;
    }
}

uintptr_t buffer[100];
int i = 0;
NSMutableString *reStr = [NSMutableString stringWithFormat:@"Stack of thread: %u:\n CPU used: %.1f percent\n user time: %d second\n", thread, threadInfoSt.cpuUsage, threadInfoSt.userTime];
#获取线程里所有栈的信息
```

通过 thread_get_state 得到 machineContext，其中包含了线程栈里所有的栈指针。

```
_STRUCT_MCONTEXT machineContext; //线程栈里所有的栈指针
//通过 thread_get_state 获取完整的 machineContext 信息，包含 thread 状态信息
mach_msg_type_number_t state_count = smThreadStateCountByCPU();
kern_return_t kr = thread_get_state(thread, smThreadStateByCPU(),
(thread_state_t)&machineContext.__ss, &state_count);
```

创建一个栈结构体来保存栈的数据。

```
//为实现通用的回溯设计结构，支持栈地址由小到大排列，地址里存储上个栈指针的地址
typedef struct SMStackFrame {
    const struct SMStackFrame *const previous;
    const uintptr_t return_address;
```

```
} SMStackFrame;

SMStackFrame stackFrame = {0};
//通过栈基址指针获取当前栈帧地址
const uintptr_t framePointer = smMachStackBasePointerByCPU(
                                &machineContext);
if (framePointer == 0 || smMemCopySafely((void *)framePointer,
&stackFrame, sizeof(stackFrame)) != KERN_SUCCESS) {
    return @"Fail frame pointer";
}
for (; i < 32; i++) {
    buffer[i] = stackFrame.return_address;
    if (buffer[i] == 0 || stackFrame.previous == 0 || smMemCopySafely(
    stackFrame.previous, &stackFrame,
    sizeof(stackFrame)) != KERN_SUCCESS) {
        break;
    }
}
#符号化
```

符号化解析的主要思想就是通过栈指针地址减去 Slide 地址得到 ASLR 偏移量。通过 ASLR 偏移量可以在 __LINKEDIT segment 中查找到字符串和符号表的位置。具体代码实现如下：

```
info->dli_fname = NULL;
info->dli_fbase = NULL;
info->dli_sname = NULL;
info->dli_saddr = NULL;
//根据地址获取相应的 image
const uint32_t idx = smDyldImageIndexFromAddress(address);
if (idx == UINT_MAX) {
    return false;
}
/*
Header
------------------
Load commands
Segment command 1 --------------|
Segment command 2              |
------------------              |
Data                            |
Section 1 data |segment 1 <----|
Section 2 data |         <----|
Section 3 data |         <----|
Section 4 data |segment 2
Section 5 data |
...            |
Section n data |
*/
/*----------Mach Header---------*/
//根据 image 的序号获取 mach_header
const struct mach_header* machHeader = _dyld_get_image_header(idx);
//返回 image_index 索引的 image 的虚拟内存地址 slide 的数量
```

```c
    //当 image_index 超出范围时返回 0
    //动态链接器加载 image 时，image 必须映射到未占用地址的进程的虚拟地址空间
    //动态链接器通过添加一个值到 image 的基地址，实现对基地址的保护
    //这个值是虚拟内存 slide 的数量
    const uintptr_t imageVMAddressSlide =
(uintptr_t)_dyld_get_image_vmaddr_slide(idx);
    /*-----------ASLR 的偏移量---------*/
    //根据 image 的 index 来获取 segment 的基地址
    //当 segment 定义 Mach-O 文件中的字节范围，以及动态链接器加载应用程序时
    //这些字节将会映射到虚拟内存中的地址和内存保护属性的地址，因此段总是虚拟内存页对齐
    //segment 包含零个或多个 section
    const uintptr_t segmentBase = smSegmentBaseOfImageIndex(idx) +
imageVMAddressSlide;
    if (segmentBase == 0) {
        return false;
    }
    //
    info->dli_fname = _dyld_get_image_name(idx);
    info->dli_fbase = (void*)machHeader;

    /*--------------Mach Segment-------------*/
    //地址最匹配的 symbol
    const nlistByCPU* bestMatch = NULL;
    uintptr_t bestDistance = ULONG_MAX;
    uintptr_t cmdPointer = smCmdFirstPointerFromMachHeader(machHeader);
    if (cmdPointer == 0) {
        return false;
    }
    //遍历每个 segment 判断目标地址是否落在该 segment 包含的范围里
    for (uint32_t iCmd = 0; iCmd < machHeader->ncmds; iCmd++) {
        const struct load_command* loadCmd = (struct load_command*)cmdPointer;
        /*----------目标 Image 的符号表---------*/
        //segment 除包含 __TEXT 和 __DATA 以外，还有 __LINKEDIT segment
        //包含动态链接器使用的原始数据，比如符号、字符串和重定位表项
        // LC_SYMTAB 描述了 __LINKEDIT segment 内查找字符串在符号表里的位置
        if (loadCmd->cmd == LC_SYMTAB) {
            //获取字符串和符号表的虚拟内存偏移量
            const struct symtab_command* symtabCmd = (struct symtab_command*)
                                                        cmdPointer;
            const nlistByCPU* symbolTable = (nlistByCPU*)(segmentBase +
                                            symtabCmd->symoff);
            const uintptr_t stringTable = segmentBase + symtabCmd->stroff;

            for (uint32_t iSym = 0; iSym < symtabCmd->nsyms; iSym++) {
                //如果 n_value 是 0，那么 symbol 指向外部对象
                if (symbolTable[iSym].n_value != 0) {
                    //给定的偏移量是文件偏移量，减去 LINKEDIT segment 的文件偏移量
                    //获得字符串和符号表的虚拟内存偏移量
                    uintptr_t symbolBase = symbolTable[iSym].n_value;
                    uintptr_t currentDistance = addressWithSlide - symbolBase;
                    //寻找最小距离 bestDistance
                    //因为 addressWithSlide 是某个方法的指令地址
                    //这个地址要大于其方法入口的地址
                    //离 addressWithSlide 越近的函数，入口越匹配
```

```
                if ((addressWithSlide >= symbolBase) && (currentDistance <=
bestDistance)) {
                    bestMatch = symbolTable + iSym;
                    bestDistance = currentDistance;
                }
            }
        }
        if (bestMatch != NULL) {
            //将虚拟内存偏移量添加到 __LINKEDIT segment 的虚拟内存地址
            //可以提供字符串和符号表的内存 address
            info->dli_saddr = (void*)(bestMatch->n_value +
                            imageVMAddressSlide);
            info->dli_sname = (char*)((intptr_t)stringTable +
                            (intptr_t)bestMatch->n_un.n_strx);
            if (*info->dli_sname == '_') {
                info->dli_sname++;
            }
            //所有 symbol 已经被处理好了
            if (info->dli_saddr == info->dli_fbase &&
                            bestMatch->n_type == 3) {
                info->dli_sname = NULL;
            }
            break;
        }
    }
    cmdPointer += loadCmd->cmdsize;
}
```

需要注意的是，使用 thread_get_state 方法会有性能消耗。由 thread_get_state 造成的性能消耗也会被监控检查出来，所以可以过滤掉包含 thread_get_state 方法的堆栈信息。

2.18.2 通过 hook 获取更多信息的方法

通过 hook 获取更多信息，比如全层级方法调用和每个方法消耗的时间，这样做的好处是什么呢？好处是可以更细化地测量时间消耗，找到耗时方法，更快地进行交互操作，使用户体验更好。

以下为可以衡量的场景：

- 响应能力
- 按钮点击
- 手势操作
- Tab 切换
- VC 的切换和转场

定下优化目标，比如滚动和动画达到 60 帧/s、在 100ms 内响应用户的操作。然后逐一检测，若有问题，则进行修复。

如何获取更多信息呢？

通过 hook objc_msgSend 方法，能够获取所有被调用的方法，方法调用层级的深度被记录后就能够得到方法调用层级之间的树状结构。通过执行前后时间的记录能够得到每个方法的耗时，这样就能获取一份完整的性能消耗信息了。

hook c 函数可以使用 FaceBook 的 fishhook。获取方法调用树状结构可以使用 InspectiveC，下面介绍具体的实现过程。

获取方法调用树结构

首先设计两个结构体 CallRecord 和 ThreadCallStack。CallRecord 记录调用方法的详细信息，包括 obj 和 SEL 等，ThreadCallStack 需要用 index 记录当前调用方法树的深度。取得 SEL 后，再通过 NSStringFromSelector 取得方法名。取得 obj 后，通过 object_getClass 得到 Class，再用 NSStringFromClass 获得类名。

```
// Shared structures.
typedef struct CallRecord_ {
//通过 object_getClass 得到 Class, 再通过 NSStringFromClass 得到类名
  id obj;
  SEL _cmd; //通过 NSStringFromSelector 方法得到方法名
  uintptr_t lr;
  int prevHitIndex;
  char isWatchHit;
} CallRecord;

typedef struct ThreadCallStack_ {
  FILE *file;
  char *spacesStr;
  CallRecord *stack;
  int allocatedLength;
  int index; //index 记录当前调用方法树的深度
  int numWatchHits;
  int lastPrintedIndex;
  int lastHitIndex;
  char isLoggingEnabled;
  char isCompleteLoggingEnabled;
} ThreadCallStack;
```

存储读取 ThreadCallStack

pthread_setspecific() 可以将私有数据设置在指定线程上，pthread_getspecific() 可以用来读取私有数据。利用这个特性就可以将 ThreadCallStack 的数据和该线程绑定在一起，随时进行数据的存取。代码如下：

```
static inline ThreadCallStack * getThreadCallStack() {
  ThreadCallStack *cs = (ThreadCallStack *)pthread_getspecific(
                    threadKey); //读取
  if (cs == NULL) {
    cs = (ThreadCallStack *)malloc(sizeof(ThreadCallStack));
#ifdef MAIN_THREAD_ONLY
    cs->file = (pthread_main_np()) ? newFileForThread() : NULL;
#else
    cs->file = newFileForThread();
#endif
    cs->isLoggingEnabled = (cs->file != NULL);
    cs->isCompleteLoggingEnabled = 0;
```

```
      cs->spacesStr = (char *)malloc(DEFAULT_CALLSTACK_DEPTH + 1);
      memset(cs->spacesStr, ' ', DEFAULT_CALLSTACK_DEPTH);
      cs->spacesStr[DEFAULT_CALLSTACK_DEPTH] = '\0';
      cs->stack = (CallRecord *)calloc(DEFAULT_CALLSTACK_DEPTH,
                  sizeof(CallRecord));  //分配 CallRecord 默认空间
      cs->allocatedLength = DEFAULT_CALLSTACK_DEPTH;
      cs->index = cs->lastPrintedIndex = cs->lastHitIndex = -1;
      cs->numWatchHits = 0;
      pthread_setspecific(threadKey, cs);  //保存数据
    }
    return cs;
}
```

记录方法调用深度

因为要记录深度，而一个方法的调用里会有很多的方法调用，所以在方法的调用中写两个方法：pushCallRecord 和 popCallRecord。pushCallRecord 在开始时调用，popCallRecord 在结束时调用。通过 pushCallRecord 方法在开始时对深度变量加 1，通过 popCallRecord 在结束时对深度变量减 1。

```
//开始时
static inline void pushCallRecord(id obj, uintptr_t lr, SEL _cmd,
    ThreadCallStack *cs) {
    int nextIndex = (++cs->index);  //增加深度值
    if (nextIndex >= cs->allocatedLength) {
        cs->allocatedLength += CALLSTACK_DEPTH_INCREMENT;
        cs->stack = (CallRecord *)realloc(cs->stack,
                    cs->allocatedLength * sizeof(CallRecord));
        cs->spacesStr = (char *)realloc(cs->spacesStr,
                        cs->allocatedLength + 1);
        memset(cs->spacesStr, ' ', cs->allocatedLength);
        cs->spacesStr[cs->allocatedLength] = '\0';
    }
    CallRecord *newRecord = &cs->stack[nextIndex];
    newRecord->obj = obj;
    newRecord->_cmd = _cmd;
    newRecord->lr = lr;
    newRecord->isWatchHit = 0;
}
//结束时
static inline CallRecord * popCallRecord(ThreadCallStack *cs) {
    return &cs->stack[cs->index--];  //减少深度值
}
```

在 objc_msgSend 前、后插入执行方法

hook objc_msgSend 需要在调用前和调用后分别加入 pushCallRecord 和 popCallRecord。因为需要在调用后的这个时机插入一个方法，并且不可能编写一个保留未知参数并跳转到 C 语言中任意函数指针的函数，那么这就需要用到汇编语言来完成了。

下面针对 ARM64 进行分析。ARM 64 有31个64 位 的整数型寄存器，用 x0 到 x30 表示。对 ARM64 进行分析的主要思路就是先入栈参数，参数寄存器是 x0 到 x7。对于

objc_msgSend 方法来说，第一个参数是传入对象，放入 x0，第二个参数是选择器 _cmd，放入 x1。syscall 的 number 会放入 x8，然后将交换寄存器中用于返回的寄存器 lr 移入 x1。先让 pushCallRecord 能够执行，再执行原始的 objc_msgSend，保存返回值，最后让 popCallRecord 能够执行。具体代码如下：

```
static void replacementObjc_msgSend() {
    __asm__ volatile (
    // sp 是堆栈寄存器，存放栈的偏移地址，每次都指向栈顶
    // 保存 {q0-q7}，偏移地址到 sp 寄存器
      "stp q6, q7, [sp, #-32]!\n"
      "stp q4, q5, [sp, #-32]!\n"
      "stp q2, q3, [sp, #-32]!\n"
      "stp q0, q1, [sp, #-32]!\n"
    // 保存 {x0-x8, lr}
      "stp x8, lr, [sp, #-16]!\n"
      "stp x6, x7, [sp, #-16]!\n"
      "stp x4, x5, [sp, #-16]!\n"
      "stp x2, x3, [sp, #-16]!\n"
      "stp x0, x1, [sp, #-16]!\n"
    // 交换参数
      "mov x2, x1\n"
      "mov x1, lr\n"
      "mov x3, sp\n"
    // 调用 preObjc_msgSend，使用 bl label 语法。bl 执行一个分支链接操作
    // label 是无条件分支的，基于当前指令的地址偏移，范围是从 -128MB 到 +128MB
      "bl __Z15preObjc_msgSendP11objc_objectmP13objc_selectorP9RegState_\n"
      "mov x9, x0\n"
      "mov x10, x1\n"
      "tst x10, x10\n"
    // 读取 {x0-x8, lr}，从保存到 sp 栈顶的偏移地址读起
      "ldp x0, x1, [sp], #16\n"
      "ldp x2, x3, [sp], #16\n"
      "ldp x4, x5, [sp], #16\n"
      "ldp x6, x7, [sp], #16\n"
      "ldp x8, lr, [sp], #16\n"
    // 读取 {q0-q7}
      "ldp q0, q1, [sp], #32\n"
      "ldp q2, q3, [sp], #32\n"
      "ldp q4, q5, [sp], #32\n"
      "ldp q6, q7, [sp], #32\n"
      "b.eq Lpassthrough\n"
    // 调用原始 objc_msgSend，使用 blr xn 语法
    // blr 除从指定寄存器读取新的 PC 值外，效果和 bl 一样
    // xn 是通用寄存器的 64 位名称分支地址，范围是从 0 到 31
      "blr x9\n"
    // 保存 {x0-x9}
      "stp x0, x1, [sp, #-16]!\n"
      "stp x2, x3, [sp, #-16]!\n"
      "stp x4, x5, [sp, #-16]!\n"
      "stp x6, x7, [sp, #-16]!\n"
      "stp x8, x9, [sp, #-16]!\n"
    // 保存 {q0-q7}
```

```
        "stp q0, q1, [sp, #-32]!\n"
        "stp q2, q3, [sp, #-32]!\n"
        "stp q4, q5, [sp, #-32]!\n"
        "stp q6, q7, [sp, #-32]!\n"
     // 调用 postObjc_msgSend hook.
        "bl __Z16postObjc_msgSendv\n"
        "mov lr, x0\n"
     // 读取 {q0-q7}
        "ldp q6, q7, [sp], #32\n"
        "ldp q4, q5, [sp], #32\n"
        "ldp q2, q3, [sp], #32\n"
        "ldp q0, q1, [sp], #32\n"
     // 读取 {x0-x9}
        "ldp x8, x9, [sp], #16\n"
        "ldp x6, x7, [sp], #16\n"
        "ldp x4, x5, [sp], #16\n"
        "ldp x2, x3, [sp], #16\n"
        "ldp x0, x1, [sp], #16\n"
        "ret\n"
        "Lpassthrough:\n"
     // br 无条件分支到寄存器中的地址
        "br x9"
        );
}
```

记录时间的方法

为了记录耗时，需要在 pushCallRecord 和 popCallRecord 里记录一下时间。下面列出了一些计算一段代码从开始到结束经历的时间的方法。

第一种：NSDate（微秒）。

```
NSDate* tmpStartData = [NSDate date];
//some code need caculate
double deltaTime = [[NSDate date] timeIntervalSinceDate:tmpStartData];
NSLog(@"cost time: %f s", deltaTime);
```

第二种：clock_t（微秒）。clock_t 表示的是占用 CPU 的时钟单元。

```
clock_t start = clock();
//some code need caculate
clock_t end = clock();
NSLog(@"cost time: %f s", (double)(end - start)/CLOCKS_PER_SEC);
```

第三种：CFAbsoluteTime（微秒）。

```
CFAbsoluteTime start = CFAbsoluteTimeGetCurrent();
//some code need caculate
CFAbsoluteTime end = CFAbsoluteTimeGetCurrent();
NSLog(@"cost time = %f s", end - start); //s
```

第四种：CFTimeInterval（纳秒）。

```
CFTimeInterval start = CACurrentMediaTime();
//some code need caculate
CFTimeInterval end = CACurrentMediaTime();
```

```
NSLog(@"cost time: %f s", end - start);
```

第五种：mach_absolute_time（纳秒）。

```
uint64_t start = mach_absolute_time();
//some code need caculate
uint64_t end = mach_absolute_time();
uint64_t elapsed = 1e-9 *(end - start);
```

最后两种方法更适合于方法耗时的记录，其他三种方法（NSDate、clock_t、CFAbsoluteTime）返回的时钟时间从时钟偏移量的角度看会和网络时间同步，而 CFTimeInterval 和 mach_absolute_time 方法是基于内建时钟的。选择任意一种方法加到 pushCallRecord 和 popCallRecord 方法里，将获取到的值相减就可以获得方法消耗的时间了。

2.18.3　hook msgsend 方法

使用 C 语言编写的方法 objc_msgSend 是如何 hook 的呢？首先，dyld 是通过更新 Mach-O 二进制的 __DATA segment 特定部分中的指针来绑定 lazy 和 non-lazy 符号的，根据确认的符号名称找出 rebind symbol 里对应替换的符号名称，通过更新两个符号位置来进行符号的重新绑定。下面针对关键代码进行分析。

遍历 dyld

遍历 dyld 里的所有 image，取出 image header 和 slide。注意第一次调用时主要注册 callback。

```
if (!_rebindings_head->next) {
    _dyld_register_func_for_add_image(_rebind_symbols_for_image);
} else {
    uint32_t c = _dyld_image_count();
    for (uint32_t i = 0; i < c; i++) {
        _rebind_symbols_for_image(_dyld_get_image_header(i),
            _dyld_get_image_vmaddr_slide(i));
    }
}
```

找出与符号表相关的 command

接下来，找到与符号表相关的 command，包括 linkedit segment command、symtab command 和 dysymtab command。方法如下：

```
segment_command_t *cur_seg_cmd;
segment_command_t *linkedit_segment = NULL;
struct symtab_command* symtab_cmd = NULL;
struct dysymtab_command* dysymtab_cmd = NULL;

uintptr_t cur = (uintptr_t)header + sizeof(mach_header_t);
for (uint i = 0; i < header->ncmds; i++, cur += cur_seg_cmd->cmdsize) {
    cur_seg_cmd = (segment_command_t *)cur;
    if (cur_seg_cmd->cmd == LC_SEGMENT_ARCH_DEPENDENT) {
        if (strcmp(cur_seg_cmd->segname, SEG_LINKEDIT) == 0) {
```

```
            linkedit_segment = cur_seg_cmd;
        }
    } else if (cur_seg_cmd->cmd == LC_SYMTAB) {
        symtab_cmd = (struct symtab_command*)cur_seg_cmd;
    } else if (cur_seg_cmd->cmd == LC_DYSYMTAB) {
        dysymtab_cmd = (struct dysymtab_command*)cur_seg_cmd;
    }
}
// Find base symbol/string table addresses
uintptr_t linkedit_base = (uintptr_t)slide + linkedit_segment->
                          vmaddr - linkedit_segment->fileoff;
nlist_t *symtab = (nlist_t *)(linkedit_base + symtab_cmd->symoff);
char *strtab = (char *)(linkedit_base + symtab_cmd->stroff);

// Get indirect symbol table (array of uint32_t indices into symbol table)
uint32_t *indirect_symtab = (uint32_t *)(linkedit_base + dysymtab_cmd->
                            indirectsymoff);
```

方法替换

有了符号表和外部输入方法替换映射表数组，就可以进行符号表访问指针地址的替换了，具体实现如下：

```
uint32_t *indirect_symbol_indices = indirect_symtab + section->reserved1;
void **indirect_symbol_bindings = (void **)((uintptr_t)slide + section->
                                  addr);
for (uint i = 0; i < section->size / sizeof(void *); i++) {
    uint32_t symtab_index = indirect_symbol_indices[i];
    if (symtab_index == INDIRECT_SYMBOL_ABS || symtab_index ==
        INDIRECT_SYMBOL_LOCAL || symtab_index == (INDIRECT_SYMBOL_LOCAL |
        INDIRECT_SYMBOL_ABS)) {
        continue;
    }
    uint32_t strtab_offset = symtab[symtab_index].n_un.n_strx;
    char *symbol_name = strtab + strtab_offset;
    if (strnlen(symbol_name, 2) < 2) {
        continue;
    }
    struct rebindings_entry *cur = rebindings;
    while (cur) {
        for (uint j = 0; j < cur->rebindings_nel; j++) {
            if (strcmp(&symbol_name[1], cur->rebindings[j].name) == 0) {
                if (cur->rebindings[j].replaced != NULL &&
                    indirect_symbol_bindings[i] != cur->rebindings[j]
                    .replacement) {
                    *(cur->rebindings[j].replaced) = indirect_symbol
                                                    _bindings[i];
                }
                indirect_symbol_bindings[i] = cur->rebindings[j]
                                              .replacement;
                goto symbol_loop;
            }
        }
        cur = cur->next;
    }
symbol_loop:;
```

2.18.4 统计方法调用频次

在某些应用场景中会发生频繁的方法调用，而有些方法的调用实际上是没有必要的。首先要将那些频繁调用的方法找出来，这样才能够更好地定位到潜在的、会造成性能浪费的方法调用。

这些频繁调用的方法要怎么找呢？大致的思路是，基于前面章节提到的记录方法调用深度的方案，保存每个调用方法的路径，对于相同路径的同一个方法，每调用一次就在数据库中对记录的次数加 1，最后做一个视图，按照调用次数的排序即可找到调用频繁的方法。下图是统计方法调用频次的结果。

接下来我们看一看具体的实现过程。

设计方法调用频次记录的结构

在先前时间消耗的 model 基础上增加路径、频次等信息。

```
@property (nonatomic, strong) NSString *className;         //类名
@property (nonatomic, strong) NSString *methodName;        //方法名
@property (nonatomic, assign) BOOL isClassMethod;          //是否是类方法
@property (nonatomic, assign) NSTimeInterval timeCost;     //时间消耗
@property (nonatomic, assign) NSUInteger callDepth;        //Call 层级
@property (nonatomic, copy)   NSString *path;              //路径
@property (nonatomic, assign) BOOL lastCall;               //是否最后一个 Call
@property (nonatomic, assign) NSUInteger frequency;        //访问频次
@property (nonatomic, strong) NSArray <SMCallTraceTimeCostModel *> *subCosts;
```

拼装方法路径

在遍历 SMCallTrace 记录的方法调用的 model 和遍历方法调用子方法调用时，将路径拼装好，记录到数据库中。

```
for (SMCallTraceTimeCostModel *model in arr) {
    //记录方法路径
```

```
        model.path = [NSString
stringWithFormat:@"[%@ %@]",model.className,model.methodName];
        [self appendRecord:model to:mStr];
    }
}

+ (void)appendRecord:(SMCallTraceTimeCostModel *)cost to:
  (NSMutableString *)mStr {
    [mStr appendFormat:@"%@\n path%@\n",[cost des],cost.path];
    if (cost.subCosts.count < 1) {
        cost.lastCall = YES;
    }
    //记录到数据库中
    [[[SMLagDB shareInstance] increaseWithClsCallModel:cost]
      subscribeNext:^(id x) {}];
    for (SMCallTraceTimeCostModel *model in cost.subCosts) {
        //记录方法的子方法的路径
        model.path = [NSString stringWithFormat:@"%@ - [%@ %@]",
           cost.path,model.className,model.methodName];
        [self appendRecord:model to:mStr];
    }
}
```

记录方法调用频次数据库

先创建数据库。lastcall 布尔值用来表示是否是最后一个方法的调用，展示时只取最后一个方法即可。因为数据库会记录完整路径，所以可以知道父方法和来源方法。

```
    _clsCallDBPath = [PATH_OF_DOCUMENT stringByAppendingPathComponent:
                   @"clsCall.sqlite"];
    if ([[NSFileManager defaultManager] fileExistsAtPath:_clsCallDBPath] ==
      NO) {
        FMDatabase *db = [FMDatabase databaseWithPath:_clsCallDBPath];
        if ([db open]) {
            /*
              cid: 主 id
              fid: 父 id 暂时不用
              cls: 类名
              mtd: 方法名
              path: 完整路径标识
              timecost:消耗时长
              calldepth: 层级
              frequency: 调用次数
              lastcall:最后一个 call
            */
            NSString *createSql = @"create table clscall (cid INTEGER PRIMARY KEY
AUTOINCREMENT NOT NULL, fid integer, cls text, mtd text, path text, timecost integer,
calldepth integer, frequency integer, lastcall integer)";
            [db executeUpdate:createSql];
        }
    }
```

添加记录时，需要先检查数据库里是否有方法名和路径都相同的记录，这样可以给 frequency 字段加 1，以达到记录频次的目的。

```
    FMResultSet *rsl = [db executeQuery:@"select cid,frequency from clscall where path
= ?", model.path];
    if ([rsl next]) {
        //有相同路径就可以更新路径访问频率
        int fq = [rsl intForColumn:@"frequency"] + 1;
        int cid = [rsl intForColumn:@"cid"];
        [db executeUpdate:@"update clscall set frequency = ? where cid = ?", @(fq),
@(cid)];
    } else {
        //没有相同路径就添加一条记录
        NSNumber *lastCall = @0;
        if (model.lastCall) {
            lastCall = @1;
        }
        [db executeUpdate:@"insert into clscall (cls, mtd, path, timecost, calldepth,
frequency, lastcall) values (?, ?, ?, ?, ?, ?, ?)", model.className, model.methodName,
model.path, @(model.timeCost), @(model.callDepth), @1, lastCall];
    }
    [db close];
    [subscriber sendCompleted];
```

检索时要注意按照调用频次字段进行排序。

```
    FMResultSet *rs = [db executeQuery:@"select * from clscall where lastcall=? order
by frequency desc limit ?, 50",@1, @(page * 50)];
    NSUInteger count = 0;
    NSMutableArray *arr = [NSMutableArray array];
    while ([rs next]) {
        SMCallTraceTimeCostModel *model = [self clsCallModelFromResultSet:rs];
        [arr addObject:model];
        count ++;
    }
    if (count > 0) {
        [subscriber sendNext:arr];
    } else {
        [subscriber sendError:nil];
    }
    [subscriber sendCompleted];
    [db close];
```

2.18.5 找出 CPU 使用的线程堆栈

在前面检测卡顿打印堆栈的讲解中，提到使用 thread_info 能够获取各个线程的 CPU 消耗情况。但是当 CPU 不在主线程时，即使消耗很大，也不一定会造成卡顿，结果导致卡顿检测无法检测出更多 CPU 消耗的情况。所以只能通过轮询监控各线程的 CPU 使用情况。这里自定义一个阈值 70%，针对这个阈值来跟踪、定位哪些方法是耗电的。下图列出了 CPU 过载时堆栈记录的结果。

有了获取线程堆栈的基础后，线程堆栈的打印实现起来就轻松多了，代码如下所示。

```objc
//轮询检查多个线程 CPU 情况
+ (void)updateCPU {
    thread_act_array_t threads;
    mach_msg_type_number_t threadCount = 0;
    const task_t thisTask = mach_task_self();
    kern_return_t kr = task_threads(thisTask, &threads, &threadCount);
    if (kr != KERN_SUCCESS) {
        return;
    }
    for (int i = 0; i < threadCount; i++) {
        thread_info_data_t threadInfo;
        thread_basic_info_t threadBaseInfo;
        mach_msg_type_number_t threadInfoCount = THREAD_INFO_MAX;
        if (thread_info((thread_act_t)threads[i], THREAD_BASIC_INFO, (thread_info_t)threadInfo, &threadInfoCount) == KERN_SUCCESS) {
            threadBaseInfo = (thread_basic_info_t)threadInfo;
            if (!(threadBaseInfo->flags & TH_FLAGS_IDLE)) {
                integer_t cpuUsage = threadBaseInfo->cpu_usage / 10;
                if (cpuUsage > 70) {
                    //CUP 消耗大于 70% 时打印和记录堆栈
                    NSString *reStr = smStackOfThread(threads[i]);
                    //记录到数据库中
                    [[[SMLagDB shareInstance] increaseWithStackString:reStr]
                        subscribeNext:^(id x) {}];
                    NSLog(@"CPU useage overload thread stack: \n%@",reStr);
                }
            }
        }
    }
}
```

        }

2.18.6 Demo

上面提到的工具已整合到先前做的 GCDFetchFeed 里了。

- 子线程检测主线程的卡顿情况，在需要开始检测的地方添加 [[SMLagMonitor shareInstance] beginMonitor]; 即可。
- 在需要检测的地方调用 [SMCallTrace start];，在需要停止的地方调用 [SMCallTrace stop] 和 [SMCallTrace save]。SMCallTrace 还可以设置最大深度和最小耗时检测用来过滤不需要看到的信息。
- 要使用方法调用频次功能，在需要开始统计的地方加上 [SMCallTrace startWithMaxDepth:3];，记录时使用 [SMCallTrace stopSaveAndClean]；在数据库中保存时间记录的同时清理内存占用，并且可以 hook VC 的 viewWillAppear 和 viewWillDisappear。SMCallTrace 会在调用 viewWillAppear 时开始记录时间，在调用 viewWillDisappear 时记录时间到数据库，同时清理一次内存。展示结果的 ViewController 是 SMClsCallViewController 这个类，调出 SMClsCallViewController 就能够看到列表结果。

2.19 dyld

生成可执行文件后，程序就可以在启动时进行动态链接了，即进行符号和地址的绑定。程序会加载所依赖的动态库，修正地址偏移。iOS 会用 ASLR 来做地址偏移，避免被攻击。首先确定 Non-Lazy Pointer 的地址，对此地址进行符号与地址的绑定，加载所有类，最后执行 load 方法和 clang attribute 的 constructor 修饰函数。

以 2.18 节中的 Mach-O 为例继续分析，每个函数、全局变量和类都是通过符号的形式来定义和使用的。当把目标文件链接成一个执行文件时，dyld 在目标文件和动态库之间对符号做解析处理。

符号表会规定动态库的符号。使用 Xcode 自带的 nm 工具可以看到 Mach-O 文件的内容。

```
xcrun nm -nm SayHi.o
                 (undefined) external _OBJC_CLASS_$_Foo
                 (undefined) external _objc_autoreleasePoolPop
                 (undefined) external _objc_autoreleasePoolPush
                 (undefined) external _objc_msgSend
0000000000000000 (__TEXT,__text) external _main
```

- OBJC_CLASS$_Foo 表示 Foo 的 OC 符号。
- (undefined) external 表示未实现非私有，如果是私有的，就是 non-external。
- external _main 表示 main() 函数，处理 0 地址。使用 nm 工具可以看到 Foo 的 Mach-O 文件的内容。

```
xcrun nm -nm Foo.o
```

```
                (undefined) external _NSLog
                (undefined) external _OBJC_CLASS_$_NSObject
                (undefined) external _OBJC_METACLASS_$_NSObject
                (undefined) external ___CFConstantStringClassReference
                (undefined) external __objc_empty_cache
0000000000000000 (__TEXT,__text) non-external -[Foo say]
0000000000000060 (__DATA,__objc_const) non-external l_OBJC_METACLASS_RO_$_Foo
00000000000000a8 (__DATA,__objc_const) non-external
l_OBJC_$_INSTANCE_METHODS_Foo
00000000000000c8 (__DATA,__objc_const) non-external l_OBJC_CLASS_RO_$_Foo
0000000000000110 (__DATA,__objc_data) external _OBJC_METACLASS_$_Foo
0000000000000138 (__DATA,__objc_data) external _OBJC_CLASS_$_Foo
```

因为 undefined 符号表示该文件类并没有实现，所以在目标文件和 Foundation framework 动态库做链接处理时，链接器会尝试解析所有的 undefined 符号。

链接器可以把动态库解析成符号。链接器会记录符号对应的动态库和路径。对比一下 a.out 符号表，看一看链接器是怎么把动态库解析成符号的。

```
xcrun nm -nm a.out
                (undefined) external _NSLog (from Foundation)
                (undefined) external _OBJC_CLASS_$_NSObject (from CoreFoundation)
                (undefined) external _OBJC_METACLASS_$_NSObject (from CoreFoundation)
                (undefined) external ___CFConstantStringClassReference (from CoreFoundation)
                (undefined) external __objc_empty_cache (from libobjc)
                (undefined) external _objc_autoreleasePoolPop (from libobjc)
                (undefined) external _objc_autoreleasePoolPush (from libobjc)
                (undefined) external _objc_msgSend (from libobjc)
                (undefined) external dyld_stub_binder (from libSystem)
0000000100000000 (__TEXT,__text) [referenced dynamically] external __mh_execute_header
0000000100000e90 (__TEXT,__text) external _main
0000000100000f10 (__TEXT,__text) non-external -[Foo say]
0000000100001130 (__DATA,__objc_data) external _OBJC_METACLASS_$_Foo
0000000100001158 (__DATA,__objc_data) external _OBJC_CLASS_$_Foo
```

有了更多的信息后，就可以知道在哪个动态库中能够找到 undefined 的符号了。

通过 otool 可以找到所需的库的位置。

```
xcrun otool -L a.out
a.out:
    /System/Library/Frameworks/Foundation.framework/Versions/C/Foundation (compatibility version 300.0.0, current version 1349.25.0)
    /usr/lib/libSystem.B.dylib (compatibility version 1.0.0, current version 1238.0.0)
    /System/Library/Frameworks/CoreFoundation.framework/Versions/A/CoreFoundation (compatibility version 150.0.0, current version 1348.28.0)
    /usr/lib/libobjc.A.dylib (compatibility version 1.0.0, current version 228.0.0)
```

libSystem 里有很多我们熟悉的库。

- libdispatch：GCD。

- libsystem_c：C 语言库。
- libsystem_blocks：Block。
- libcommonCrypto：加密，比如 MD 5。

dylib 这种格式表示的是动态库，编译的时候动态库不会被编译到执行文件中，在程序执行的时候才进行链接。这样动态库就不用算到包的大小里，并且可以不更新执行程序就更新库。

下面代码打印出了 a.out 加载的库。

```
(export DYLD_PRINT_LIBRARIES=; ./a.out )
dyld: loaded: /Users/didi/Downloads/./a.out
dyld: loaded: /System/Library/Frameworks/Foundation.framework/Versions/C/Foundation
dyld: loaded: /usr/lib/libSystem.B.dylib
dyld: loaded: /System/Library/Frameworks/CoreFoundation.framework/Versions/A/CoreFoundation
...
```

因为 Foundation 还会依赖一些其他的动态库，其他的库还会再依赖更多的库，这样相互依赖的符号会很多，所以处理的时间会比较长。此时，系统的动态链接器会使用共享缓存。共享缓存在 /var/db/dyld/中。当加载 Mach-O 文件时，动态链接器会先检查是否有共享内存。每个进程都会在自己的地址空间中映射动态库使用的共享缓存，这样可以优化启动速度。

dyld 的作用如下：

- kernel 做启动程序初始准备时，由 dyld 负责。
- 启动程序本身是基于非常简单的原始堆栈设置的内核进程。
- dyld使用共享缓存来处理递归依赖带来的性能问题。ImageLoader 会读取二进制文件，其中包含了类、方法等各种符号。
- 绑定 non-lazy 的符号并设置用于 lazy bind 的必要表，将这些库链接到执行文件。
- 为可执行文件运行静态初始化。
- 设置可执行文件的 main 函数的参数并调用 main 函数。
- 在执行期间，通过绑定符号处理，对 lazily-bound 符号存根的调用提供 runtime 动态加载服务（通过 dl*() 这个 API），并为 gdb 和其他调试器提供钩子以获得关键信息。runtime 会调用 map_images 做解析、处理，load_images 调用 call_load_methods 方法遍历所有加载了的 class，按照继承层级依次调用 +load 方法。
- 在 main 函数返回后，运行 static terminator。
- 在某些情况下，一旦 main 函数返回，就需要调用 libSystem 的 _exit。

通过 RetVal 的可调试的 objc/runtime 库来设置断点，查看调用的 runtime 方法及实现，还可以查看运行时如何调用 map_images，以及如何调用 +load 方法。在 debug-objc 下创建一个类，在 +load 方法里，设置断点查看+load 的调用堆栈，代码如下：

```
0 +[someclass load]
1 call_class_loads()
2 ::call_load_methods
3 ::load_images(const char *path __unused, const struct mach_header *mh)
4 dyld::notifySingle(dyld_image_states, ImageLoader const*,
```

```
ImageLoader::InitializerTimingList*)
    11 _dyld_start
```

在 load_images 方法里设置断点，执行 p path 命令可以打印出所有加载的动态库，这个方法的 hasLoadMethods 可用于快速判断是否有 +load 方法。

prepare_load_methods 方法会获取所有类的列表，然后收集其中的 +load 方法。查看代码，可以发现 class 的 +load 方法是先执行的，然后执行 Category。为什么这样做？原因可以通过 prepare_load_methods 看出。在遍历 class 的 +load 方法时 prepare_load_methods 会执行 schedule_class_load 方法，schedule_class_load 方法会递归到根节点来满足 class 收集完整语法树的需求。

最后，call_load_methods 会创建一个 autoreleasePool，使用函数指针来动态调用类和 Category 的 +load 方法。

如果想了解 Cocoa 的 Foundation 库，可以通过 [GNUStcp]源代码来学习。比如，想了解 NSNotificationCenter 是按什么顺序发送通知的，可以查看 NSNotificationCenter.m 里的 addObserver 方法和 postNotification 方法，看看观察者是怎么被添加的及怎么被遍历通知到的。

dyld 是开源的，可以访问 Github 网站，在 Apple Open Source 中查看 dyld 相关代码和文档。

另外，可以看一看苹果公司的 Developer 网站上的 *Optimizing App Startup Time*（WWDC 2016 Session 406）视频，视频里会对程序开始启动到 main 函数之间的 dyld 内部细节进行详细讲解，并提供构建代码的最佳实践以优化启动时间。

2.20 LLVM 工具链

本节将详细介绍如何安装、编译 LLVM 的源代码，并对 LLVM 的源代码目录做详细说明。

2.20.1 获取 LLVM

获取 LLVM 的代码如下：

```
#先下载 LLVM
svn co http://llvm.org/svn/llvm-project/llvm/trunk llvm

#在 LLVM 的 tools 目录下下载 clang
cd llvm/tools
svn co http://llvm.org/svn/llvm-project/cfe/trunk clang

#在 LLVM 的 projects 目录下下载 compiler-rt, libcxx, libcxxabi
cd ../projects
svn co http://llvm.org/svn/llvm-project/compiler-rt/trunk compiler-rt
svn co http://llvm.org/svn/llvm-project/libcxx/trunk libcxx
svn co http://llvm.org/svn/llvm-project/libcxxabi/trunk libcxxabi

#在 clang 的 tools 下安装 extra 工具
```

```
cd ../tools/clang/tools
svn co http://llvm.org/svn/llvm-project/clang-tools-extra/trunk extra
```

2.20.2 编译 LLVM 的源代码

编译 LLVM 的源代码如下：

```
brew install gcc
brew install cmake
mkdir build
cd build
cmake -DCMAKE_BUILD_TYPE=Release -DBUILD_SHARED_LIBS=ON -DLLVM_TARGETS_TO_BUILD="AArch64;X86" -G "Unix Makefiles" ..
make j8
#安装
make install
#如果找不到标准库, Xcode 需要安装 xcode-select --install
```

安装后的目录结构如下图所示。

如果需要在 Xcode 环境进行编译，可以使用下面的命令。

```
#如果希望以 xcodeproject 方式构建，可以使用 -GXcode
mkdir xcodeBuild
cd xcodeBuild
cmake -GXcode /path/to/llvm/source
```

Xcode 可编译项目的目录结构如下图所示。

在 bin 下存放着工具链，有了这些工具链我们就能够完成源代码的编译了。编译出来的工具链如下图所示。

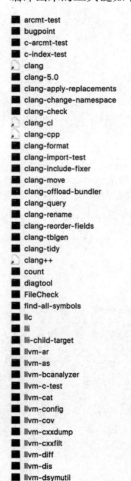

2.20.3 LLVM 源代码工程目录介绍

- llvm/examples/：LLVM IR 和 JIT 应用示例。
- llvm/include/：导出的头文件。
- llvm/lib/：主要源文件都在这里。
- llvm/project/：创建的基于 LLVM 的项目。
- llvm/test/：基于 LLVM 的回归测试，健全检查。
- llvm/suite/：性能和基准测试套件。
- llvm/tools/：基于 lib 构建的可执行文件，用户通过这些程序进行交互，使用-help 可以查看各个工具的详细使用说明。
- llvm/utils/：LLVM 源代码的实用工具，比如查找 LLC 和 LLI 生成代码差异的工具、Vim 或 Emacs 的语法高亮工具等。

2.20.4 lib 目录介绍

- llvm/lib/IR/：核心类，比如 Instruction 和 BasicBlock。
- llvm/lib/AsmParser/：汇编语言解析器。
- llvm/lib/Bitcode/：读取和写入字节码。
- llvm/lib/Analysis/：各种对程序的分析，比如 Call Graphs、Induction Variables、Natural Loop Identification，等等。
- llvm/lib/Transforms/： IR-to-IR 程序的变换。
- llvm/lib/Target/：对目标机器指令的描述，比如 x86 架构机器指令的描述。
- llvm/lib/CodeGen/：主要包括代码生成、指令选择器、指令调度和寄存器分配。
- llvm/lib/ExecutionEngine/：在解释执行和 JIT 编译场景能够直接在运行时执行字节码的库。

2.20.5 工具链命令介绍

基本命令

- llvm-as：汇编器，将 .ll 汇编成字节码。
- llvm-dis：反汇编器，将字节码编成可读的 .ll 文件。
- opt：字节码优化器。
- llc：静态编译器，将字节码编译成汇编代码。
- lli：直接执行 LLVM 字节码。
- llvm-link：字节码链接器，可以把多个字节码文件链接成一个。
- llvm-ar：字节码文件打包器。
- llvm-lib：LLVM lib.exe 兼容库工具。
- llvm-nm：列出字节码和符号表。
- llvm-config：打印 LLVM 编译选项。

- llvm-diff：对两份源代码进行比较。
- llvm-cov：输出 coverage infomation。
- llvm-profdata：Profile 数据工具。
- llvm-stress：生成随机 .ll 文件。
- llvm-symbolizer：地址对应源代码位置，定位错误。
- llvm-dwarfdump：打印 DWARF。

调试工具

- bugpoint：自动测试案例工具。
- llvm-extract：从一个 LLVM 模块里提取一个函数。
- llvm-bcanalyzer：LLVM 字节码分析器。

开发工具

- FileCheck：灵活的模式匹配文件验证器。
- tblgen：C++ 代码生成器。
- lit：LLVM 集成测试器。
- llvm-build：LLVM 构建工程时需要的工具。
- llvm-readobj：LLVM Object 结构查看器。

第 3 章
iOS 中的大前端技术

3.1 大前端技术简介

自 iOS 和 Android 争霸时代开始,"跨平台"就是很多开发者的目标。从技术角度来看,跨平台技术之间主要的区别在于所使用的虚拟机不同,即使用内置的或自己编译的 Web 标准虚拟机,还是使用其他的自定义语言的虚拟机。对于渲染和事件处理,同样也是因为所使用的虚拟机不同而有所不同,比如基于 Web 标准的 WebKit 里面除了有虚拟机,还能对渲染和事件等进行封装处理。技术选型总是在易用和性能之间做取舍,系统内置的 Web 标准虚拟机主要用在浏览器上,以兼容历史悠久的 Web 代码,但在解析上无法实现着性能。另外,由于 CSS 设计得过于复杂、灵活,因此造成不可控的计算消耗。负责和 C++ 层打交道的 DOM 接口少,因而无法有针对性地进行优化。即使是这样,如果考虑到动态化、包大小和技术栈生态,那么 Web 标准虚拟机依然是不二选择,所以造成了现在"看起来仅仅使用 Web 技术,就可以实现跨平台"的所谓"大前端时代"。着眼于未来,使用更加精简、优雅的 WebKit 是不可逆的趋势,是等待它的到来还是参与进去?解决这种疑惑的办法是"静下心,去了解里面的核心技术—— JavaScriptCore 和 WebCore"。Flutter 就是顺应这个潮流的一种尝试,是精简到只有一些必要功能的虚拟机和渲染引擎。但其在构建技术生态上能否成功还有待观望,Flutter 的底层技术类似 WebKit,可以被看成是一个专门针对性能优化而删减了的 WebKit。细心的读者会发现,大前端依赖的不是前端语言或者前端的各种编程框架,而是虚拟机和渲染引擎。所以 3.2 节从 Weex 入手开始介绍。React Native 使用的技术类似于 Weex。React Native 和 Weex 只是在使用的 JavaScript 框架和实现的细节上有所差别。从后期的发展来看,类似 Weex 的技术最终会将不同的渲染引擎统一成一个渲染引擎,这个渲染引擎类似 Web Core。3.3 节和 3.4 节将分别详细介绍 JavaScriptCore 和 WebCore。

3.2 Weex 实现技术

3.2.1 将 iOS 工程集成 WeexSDK

先创建一个 iOS Demo 工程,通过 CocoaPod 集成 WeexSDK。

```
# platform :ios, '8.0'
use_frameworks!
target 'WeexiOSDemo' do
    pod 'WeexSDK'
```

```
end
```

在 AppDelegate 里引入 Weex SDK。

```
#import <WeexSDK/WXAppConfiguration.h>
#import <WeexSDK/WXSDKEngine.h>

...

- (BOOL)application:(UIApplication *)application
didFinishLaunchingWithOptions:(NSDictionary *)launchOptions {
    // Override point for customization after application launch.
    //business configuration
    [WXAppConfiguration setAppGroup:@"Starming"];
    [WXAppConfiguration setAppName:@"TestDemo"];
    [WXAppConfiguration setAppVersion:@"1.0.0"];

    //init sdk environment
    [WXSDKEngine initSDKEnvironment];

    [WXSDKEngine registerModule:@"test" withClass:NSClassFromString
      (@"WXTestModule")];

    return YES;
}
```

下面为在 ViewController 里引入 Weex 的实例。

```
#import "ViewController.h"
#import <WeexSDK/WXSDKInstance.h>

@interface ViewController ()

@property (nonatomic, strong) WXSDKInstance *instance;
@property (nonatomic, strong) UIView *weexView;

@end

@implementation ViewController

- (void)viewDidLoad {
    [super viewDidLoad];

    _instance = [[WXSDKInstance alloc] init];
    _instance.viewController = self;
    _instance.frame = self.view.frame;

    __weak typeof(self) weakSelf = self;
    _instance.onCreate = ^(UIView *view) {
        [weakSelf.weexView removeFromSuperview];
        weakSelf.weexView = view;
        [weakSelf.view addSubview:weakSelf.weexView];
    };

    _instance.onFailed = ^(NSError *error) {
```

```
        //process failure
    };
    
    _instance.renderFinish = ^ (UIView *view) {
        //process renderFinish
    };
    
    NSString * entryURL = @"bundlejs/index.js";
    NSURL *url = [NSURL URLWithString:[NSString stringWithFormat:@"%@%@",
                 [[NSBundle bundleForClass:self] resourceURL].
                 absoluteString, entryURL]];
    
    [_instance renderWithURL:url options:@{@"bundleUrl":[url absoluteString]}
data:nil];
}

- (void)dealloc
{
    [_instance destroyInstance];
}
```

3.2.2 自定义端内能力的 Module

新建一个遵循 WXModuleProtocol 协议的类。

```
#import <Foundation/Foundation.h>
#import "WXModuleProtocol.h"

@interface WXTestModule : NSObject <WXModuleProtocol>

@end
```

实现这个类，同时将方法暴露给 Weex 模板的 JavaScript 调用。

```
#import "WXTestModule.h"

@implementation WXTestModule

WX_EXPORT_METHOD(@selector(showSomthing:))

- (void)showSomthing:(NSString *)input {
    if (!input) {
        return;
    }
    NSLog(@"%@", input);
}

@end
```

再对这个类进行注册。

```
[WXSDKEngine registerModule:@"test"
withClass:NSClassFromString(@"WXTestModule")];
```

接下来，这个方法在 Weex 模板里可以被直接调用了。

```
weex.requireModule("test").showSomthing("It's time to show something")
```

看到这里,你一定会好奇 Weex 是怎么使这些自定义的方法能够让 JavaScript 在运行时调用的?我们从自定义类里的宏 WX_EXPORT_METHOD 来着手分析。先把相关的宏都列出来,如下所示。

```
#define WX_CONCAT(a, b)   a ## b

#define WX_CONCAT_WRAPPER(a, b)    WX_CONCAT(a, b)

#define WX_EXPORT_METHOD_INTERNAL(method, token) \
+ (NSString *)WX_CONCAT_WRAPPER(token, __LINE__) { \
    return NSStringFromSelector(method); \
}

#define WX_EXPORT_METHOD(method) WX_EXPORT_METHOD_INTERNAL
    (method,wx_export_method_)
```

可以看出来,这里定义了一个将 wx_export_method_ 和行号组成一个便于查找且不会重名的方法,方便程序在运行时在方法列表里进行遍历、查找和整理。接下来就能以 wx_export_method_ 前缀作为引子,在 WeexSDK 的 WXInvocationConfig 类里找到 registerMethods。

```
- (void)registerMethods
{
    Class currentClass = NSClassFromString(_clazz);

    if (!currentClass) {
        WXLogWarning(@"The module class [%@] doesn't exit! ", _clazz);
        return;
    }

    while (currentClass != [NSObject class]) {
        unsigned int methodCount = 0;
        Method *methodList = class_copyMethodList(object_getClass(
                            currentClass), &methodCount);
        for (unsigned int i = 0; i < methodCount; i++) {
            NSString *selStr = [NSString stringWithCString:sel_getName(
                            method_getName(methodList[i])) encoding:
                            NSUTF8StringEncoding];
            BOOL isSyncMethod = NO;
            if ([selStr hasPrefix:@"wx_export_method_sync_"]) {
                isSyncMethod = YES;
            } else if ([selStr hasPrefix:@"wx_export_method_"]) {
                isSyncMethod = NO;
            } else {
                continue;
            }

            NSString *name = nil, *method = nil;
            SEL selector = NSSelectorFromString(selStr);
            if ([currentClass respondsToSelector:selector]) {
                method = ((NSString* (*)(id, SEL))[currentClass
```

```
                    methodForSelector:selector])(currentClass,
                    selector);
            }

            if (method.length <= 0) {
                WXLogWarning(@"The module class [%@] doesn't has any method! ",
_clazz);
                continue;
            }

            NSRange range = [method rangeOfString:@":"];
            if (range.location != NSNotFound) {
                name = [method substringToIndex:range.location];
            } else {
                name = method;
            }

            NSMutableDictionary *methods = isSyncMethod ? _syncMethods :
_asyncMethods;
            [methods setObject:method forKey:name];
        }

        free(methodList);
        currentClass = class_getSuperclass(currentClass);
    }
}
```

registerMethods 会通过类名 clazz，使用 class_copyMethodList 得到类里的方法集合。前面定义的 wx_export_method_ 这个前缀会在遍历集合时起到作用，遍历时发现满足这个前缀的就都放到一个字典里保存起来。asyncMethods 字典的作用是保存 wx_export_method_sync 前缀。这个前缀和 wx_export_method 的区别是什么呢？由于 WX_EXPORT_METHOD 都是异步的，所以如果想要同步获取回调结果，可以在提供原生方法时使用 WX_EXPORT_METHOD_SYNC 宏。这个宏定义的前缀就是 wx_export_method_sync_。

调用 WXInvocationConfig 的 registerMethods。registerMethods 是在 WXModule 类的 registerModule 中调用的。代码如下：

```
- (NSString *)_registerModule:(NSString *)name withClass:(Class)clazz
{
    WXAssert(name && clazz, @"Fail to register the module, please check if the
parameters are correct！");

    [_moduleLock lock];
    //allow to register module with the same name;
    WXModuleConfig *config = [[WXModuleConfig alloc] init];
    config.name = name;
    config.clazz = NSStringFromClass(clazz);
    [config registerMethods];
    [_moduleMap setValue:config forKey:name];
    [_moduleLock unlock];
```

```
    return name;
}
```

下面是注册自定义类的方法。

```
[WXSDKEngine registerModule:@"test"
withClass:NSClassFromString(@"WXTestModule")];
```

每个自定义的类注册完成后，就被字典记录维护起来了。

3.2.3　读取用 JavaScript 写的 Weex 内容

虽然我们已经将 Weex 集成到了工程中，并且开发了自定义的 Native 的功能供 JavaScript 调用，但运行起来还是什么都没有。造成这种情况的原因是我们还没有写 Weex 的 JavaScript 代码，也没有页面模板。

要用 Weex 开发程序，需要先搭建一个 Weex 的开发环境。具体步骤可以在 Weex 官网查看。通过 Weex 的脚手架 weex create awesome-app 创建一个模板项目，需要开发的内容在 src 目录下。Weex 支持 Vue 开发，也支持自己定制一个其他的 JavaScript 库。build 会生成一个 bundle，存在 dist 目录下。生成的 bundle 可以采用服务器下发的方式，让端渲染可以直接集成到 App 端内的 bundle 里。我们来看看端内效果。下面是调用端内 Weex bundle 的代码。

```
NSString * entryURL = @"bundlejs/index.js";
NSURL *url = [NSURL URLWithString:[NSString
stringWithFormat:@"%@%@",[[NSBundlebundleForClass:self]
resourceURL].absoluteString, entryURL]];

[_instance renderWithURL:url options:@{@"bundleUrl":[url absoluteString]}
data:nil];
```

还记得前面为 JavaScript 准备的 Module 吗？现在我们可以到 Vue 模板的 script 标签里去调用这个方法，代码如下所示。

```
<script>
import HelloWorld from './components/HelloWorld.vue'
export default {
  name: 'App',
  components: {
    HelloWorld
  },
  data () {
    return {
      logo: 'https://gw.alicdn.com/tfs/
             TB1yopEdgoQMeJjy1XaXXcSsFXa-640-302.png'
    }
  }
}
weex.requireModule("test").showSomthing("It's time to show something")
</script>
```

从编译工程可以看到，控制台输出了我们自己创建的命名为 WXTestModule 的 Weex Module 里的 showSomething 输出的日志，输出的内容正好是我们在 JavaScript 传入的参数内容 "It's time to show something"。

3.2.4 从 Vue 代码到 JS bundle

Weex 的 .we 文件使用的是自己设计的 DSL 模板语法，类似于 HTML，也支持样式和 JavaScript，最终会被解析成 JavaScript 可识别的对象，也就是 JS bundle。解析.we 文件使用的是 weex-loader 库。在解析 HTML 标签时，使用的是第三方开源的 npm 组件 parse 5，会将 HTML 标签解析后生成 JSON 对象。生成的 JSON 对象可以通过官方 playground 对比查看。

weex-loader 先通过 parse 5 解析 .we 文件，将其解析成 JSON Object。对应的 Template 和 Script 等标签都会有对应的库去解析、处理，如下图所示。

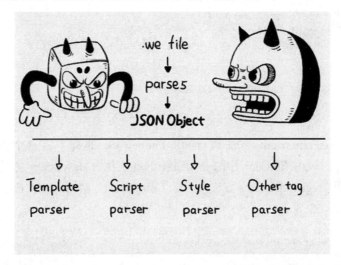

整个处理过程可以查看 weex-loader/lib/loader.js 。不同类型的标签由不同的组件处理，比如样式标签是由 weex-styler 处理的，它主要处理 Weex 所支持的 CSS 样式，由于 Weex 并没有支持 CSS 所有类型的标签，这样就需要专门对解析后获得的 JSON Object 类型的数据进行遍历校验。这里会用到另一个第三方 CSS 组件。

Template 类型的标签对应的处理组件是 weex-templater，其处理的主要是 exp 模块，代码在 /weex-templater/lib/exp.js 里。

3.2.5 在端内运行 JS bundle 的原理

在端内会通过 JS Framework 来解析生成好的 JS bundle。解析的结果是 JSON 格式的 Virtual DOM。JS Framework 会集成在 WeexSDK 里，叫作 native-bundle-main.js。这个文件是没法直接读取的，需要在 Weex 的源代码里查看。JS Framework 会在 WeexSDK 初始化的时候执行。

```objc
+ (void)initSDKEnvironment:(NSString *)script
{
    WX_MONITOR_PERF_START(WXPTInitalize)
    WX_MONITOR_PERF_START(WXPTInitalizeSync)

    if (!script || script.length <= 0) {
        NSMutableString *errMsg = [NSMutableString stringWithFormat:
        @"[WX_KEY_EXCEPTION_SDK_INIT_JSFM_INIT_FAILED] script don't
        exist:%@",script];
        [WXExceptionUtils commitCriticalExceptionRT:
        @"WX_KEY_EXCEPTION_SDK_INIT" errCode:[NSString stringWithFormat:
        @"%d", WX_KEY_EXCEPTION_SDK_INIT] function:@"initSDKEnvironment"
        exception:errMsg extParams:nil];
        WX_MONITOR_FAIL(WXMTJSFramework, WX_ERR_JSFRAMEWORK_LOAD, errMsg);
        return;
    }
    static dispatch_once_t onceToken;
    dispatch_once(&onceToken, ^{
        [self registerDefaults];
        [[WXSDKManager bridgeMgr] executeJsFramework:script];
    });

    WX_MONITOR_PERF_END(WXPTInitalizeSync)

}
```

使用 executeJsFramework 方法执行 JS Framework 的脚本，这个方法最终会通过 callJSMethod 再到 JavaScriptCore 中的 invokeMethod 来执行 JavaScript 方法。WeexSDK 在初始化时还会使用 registerDefaults 方法去注册 Component、Module 和 Handle。

```objc
+ (void)registerDefaults
{
    static dispatch_once_t onceToken;
    dispatch_once(&onceToken, ^{
        [self _registerDefaultComponents];
        [self _registerDefaultModules];
        [self _registerDefaultHandlers];
    });
}
```

那么我们自己写的 JS bundle 的脚本是怎么执行的呢？执行本地 JavaScript 脚本和执行 JS Framework 脚本在本质上是一样的，都是通过 JavaScriptCore 来执行的。如果加载使用的是 WXResourceLoader（见下图），那么它会判断其是本地的 bundle 还是网络的 bundle，然后进行不同处理。在 onFinished 里处理加载脚本后需要做的事情是在 onFinishded 这个 block 里会调用 _renderWithMainBundleString 方法。_renderWithMainBundleString 方法会调用 WXSDKManager 的 createInstance 方法来创建一个实例。创建实例主要就是调用 callJSMethod 也就是 JavaScriptCore 中的 invokeMethod。调用 callJSMethod 会调用对应的原生 UIView 方法，而对应的原生 UIView 方法是那些已经在 WeexSDK 里写好了的方法。

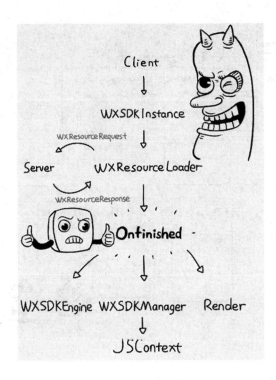

3.3 JavaScriptCore

3.3.1 JavaScriptCore 介绍

JavaScriptCore 是 JavaScript 引擎，通常会被叫作虚拟机，是专门设计来解释和执行 JavaScript 代码的。最开始的 JavaScriptCore 是从 KJS（KDE 的 JavaScript 引擎）及 PCRE 正则表达式的基础上开发的，是基于抽象语法树的解释器。在 2008 年，JavaScript 被重写了，被命名为 SquirrelFish，后来改为 SquirrelFish Extreme，又叫 Nitro。目前，其他 JavaScript 引擎还有 Google 的 V8 和 Mozilla 的 SpiderMonkey。

JavaScriptCore 能够在 Objective-C 程序中执行 JavaScript 代码，并且可以在 JavaScript 环境中插入自定义对象。

3.3.2 JavaScriptCore 全貌

下面是解析、编译 JavaScript 源代码的详细流程图。

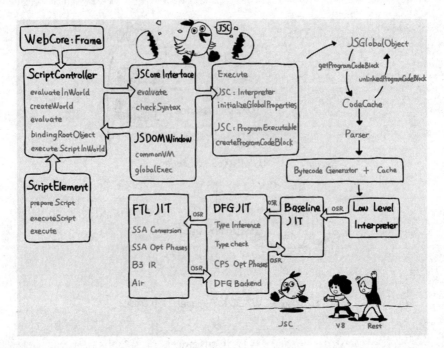

主要模块

- Lexer：词法分析器，生成 token，大部分代码都放在 parser/Lexer.cpp 文件里并在 ScriptElement 中调用、执行。
- Parser：语法分析，基于 Lexer 的 token 生成语法树。recusive descent parser 递归下降解析器，代码主要放在 parser/Parser.cpp 文件里。
- LLInt: Low Level Interpreter 执行 Parser 生成的 Bytecode。代码在 llint/ 中时，使用汇编语言编写；代码在 offlineasm/ 中时，可以编译为 x86、ARMv7 的汇编语言及 C 语言代码。LLInt 希望达成除词法和语法分析以外零启动消耗，同时遵守 JIT 调用、堆栈和寄存器的约定。
- Baseline JIT：实时编译。当程序运行性能不好时，可以使用 Baseline JIT，其在函数调用 6 次，或者某段代码循环次数大于 100 时会被触发。BaseLine JIT 的代码在 jit/ 里。BaseLine JIT 还对几乎所有堆的访问执行复杂的多态内联高速缓存。多态内联缓存是 Smalltalk 社区优化动态分发的一项经典技术。
- DFG JIT：低延迟优化 JIT。当程序运行性能差时，就用 DFG JIT 生成优化的机器码来执行。DFG JIT 在函数被调用了 60 次或者代码被循环了 1000 次时会被触发。在 LLInt（Low level Interpreter）和 Baseline JIT 中会收集一些最近调用的参数、堆，以及返回的数据等轻量级的性能信息，以方便 DFG 进行类型判断。先获取类型信息可以减少大量的类型检查工作，推测失败的 DFG 会被取消优化，也叫 OSR exit。取消优化可以是同步的也可以是异步的。取消优化后 DFG JIT 会回到 Baseline JIT，回退一定次数后 DFG JIT 会进行重新优化，收集更多统计信息，回退次数越多，再次调用 DFG 的时间间隔也会越长。重新优化使用的是指数式回退策略，以应对一些怪异的代码。DFG 代码放在 dfg/里。

- FTL：高吞吐量优化 JIT，全称 Faster Than Light，即使用 DFG 的上层优化配合 B3 的底层优化。以前全称是 Fourth Tier LLVM，底层优化使用的是 LLVM。B3 对 LLVM 做了裁剪，对 JavaScriptCore 做了特性处理，B3 IR 的接口和 LLVM IR 的接口类似。使用 B3 替换 LLVM 主要是为了减少内存开销，LLVM 主要是针对编译器。编译器在内存开销方面优化的动力没有 JIT 的高。B3 IR 将指针改成了更紧凑的整数来表示引用关系。常用的不可变的信息使用固定大小的结构表示，把可变长的信息和不常用的信息放到 B3 外部。将紧凑的数据放到数组中，使用更多的数组、更少的链表，这样形成的 IR 更省内存。Filip Pizlo 主导了这个改动，DFG JIT 也是 Filip Pizlo 编写的，为了能够更多地减少内存开销，Filip Pizlo 利用在 DFG 里已经做的 InsertionSet 将 LLVM IR 里的 def-use 去掉了，大致思路是把单向链表里批量插入的新 IR 节点先放到 InsertionSet 里，在下次遍历 IR 时再批量插入。Filip Pizlo 还把 DFG 里的 UpsilonValue 替代为 LLVM SSA 组成部分。未来 LLVM 的寄存器分配算法 Greedy 会用到 B3 中。

LLInt、Baseline JIT、DFG JIT 执行速度的对比如下图所示。

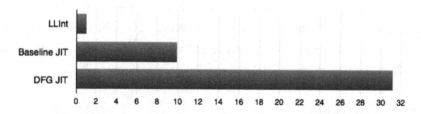

更多的说明可以参看 WebKit 官网 JavaScriptCore 的相关部分。

主要的源代码目录结构

- API：JavaScriptCore 对外的接口类。
- Assembler：不同的 CPU 会生成不同的汇编语言比如 ARM 和 x86。
- B3：FTL 里的 Backend。
- Bytecode：字节码的内容，比如类型和计算过程。
- Bytecompiler：编译字节码。
- Configuration：Xcode 的相关配置。
- Debugger：用于测试脚本的程序。
- DFG：DFG JIT 编译器。
- Disassembler：反汇编语言。
- Heap：运行时的堆和垃圾回收机制。
- FTL：第四层编译。
- Interpreter：解释器，负责解析执行 Bytecode。
- JIT：在运行时将 Bytecode 转成机器码，以便动态及时编译。
- LLInt：编译四层里的第一层，负责解释执行低效字节码。

- Parser：词法语法分析，构建语法树。
- Profiler：信息收集，能收集函数调用频率和消耗时间。
- Runtime：包含运行时对于 JavaScript 的全套操作。
- WASM：对 WebAssembly 的实现。
- YARR：运行时正则表达式的解析。

3.3.3　JavaScriptCore 与 WebCore

ScriptController 会调用 JavaScriptCore 的 evaluate 和 checkSyntax 两个接口。DOM 节点的 JSBindings 通过回溯到 JSC::JSNonFinalObject，实现和 JavaScriptCore 的绑定。

VM 是 JavaScript 的 Runtime 环境。GlobalObject 是全局对象，负责管理执行环境和 Object。ExecState 是执行脚本的对象，由 GlobalObject 管理，负责记录脚本执行上下文。

接口执行脚本，创建一个要执行的脚本的对象，这个对象是 ProgramExecutable。ProgramExecutable 负责将脚本编译成字节码，调用 Interpreter 执行字节码。

Binding 是 WebCore 为 JavaScriptCore 提供的封装层。WebKit 参照的是 W3C Web IDL。使用 IDL 定义接口规范，WebKit 使用一组 Perl 脚本来转换 IDL，初始脚本是 generate-binding.pl。通过 JSNode 生成接口，与 DOM 组件关联。执行脚本和 Frame 的 setDocument 会更新 document 对象到 JavaScriptCore 中，通过 JSDomWindowBase::updateDocument 更新到 JSC::Heap 里。

3.3.4　词法、语法分析

JavaScriptCore 的词法和语法分析程序是 JavaScriptCore 专有的。词法分析会把文本理解成很多 token，比如 a = 5; 就会被识别成下面这些 token：VARIABLE EQUAL CONSTANT END。通过语法分析可以输出语法树。

还有些其他的词法、语法分析工具，比如 Esprima 提供了一个 Demo 可以在线进行词法、语法分析。

笔者在 HTN 项目中做了 JavaScript 的 AST 的 builder，在分词和生成语法树时使用状态机处理递归下降。

WebKit 的性能目录里有用 ES6 标准实现的 ECMA-55 BASIC 词法、语法分析的测试。程序会被表示成树，每个节点都有与其相关的代码。这些代码都可以递归地调用树中的子节点。Basic 节点的结构如下。

```
{evaluate: Basic.NumberPow, left: primary, right: parsePrimary()}
```

Basic 是以不常见的方式使用生成器的，比如使用多个生成器函数，调用 yield point 和递归生成器。Basic 也有 for-of、Class、Map 和 WeakMap 功能。以上测试 ES6 的 JavaScript 写的程序可以帮助我们理解 ES6 的解析过程。

3.3.5 从代码到 JIT 的过程

ProgramExecutable 的初始化会生成 Lexer、Parser 和 Bytecode。ProgramExecutable 的入口是从 JS Binding 里调用 ScriptController::evaluateInWorld，这个方法里的参数 sourceCode 就是 JavaScript 的代码来源。在方法内部会调用 runtime/Completion.cpp 文件的 evaluate 方法，在 evaluate 方法里会调用 executeProgram 方法。总体来说，ProgramExecutable 主要是把代码编译成字节码，Interpreter 执行字节码。

```
JSValue result = vm.interpreter->executeProgram(source, exec, thisObj);
```

executeProgram 方法将源代码生成 ProgramExecutable 对象。

```
ProgramExecutable* program = ProgramExecutable::create(callFrame, source);
```

这个对象里有 StringView 对象，这样 program->source().view()就可以对源代码进行操作了。在进行操作之前 ProgramExecutable 对象会先判断这段 JavaScript 代码是否仅仅只是一个 JSON 对象。如果是，就当作 JSON 对象进行处理；如果不是，就当作普通的 JavaScript 代码进行处理。

当作普通的 JavaScript 代码进行处理时，源代码会先被编译成字节码，第一步是初始化全局属性。

```
JSObject* error = program->initializeGlobalProperties(vm, callFrame, scope);
```

处理的结果会记录在 callFrame 里。主要通过 JSGlobalObject 的 addFunction 和 addVar 方法，记录 Parser 中那些在全局空间的 Let 的全局属性、Const 的全局属性和 Class 的全局属性，或者 Var 的全局变量、Let 的全局变量和 Const 的全局变量。

ProgramExecutable 接下来会创建一个 CodeBlock。

```
ProgramCodeBlock* codeBlock;
    {
        CodeBlock* tempCodeBlock;
        JSObject* error = program->prepareForExecution<ProgramExecutable>
                    (vm, nullptr, scope, CodeForCall, tempCodeBlock);
        EXCEPTION_ASSERT(throwScope.exception() == reinterpret_cast
                                            <Exception*>(error));
        if (UNLIKELY(error))
           return checkedReturn(error);
        codeBlock = jsCast<ProgramCodeBlock*>(tempCodeBlock);
    }
```

此时，ProgramExecutable 会通过 prepareForExecution 方法，调用 prepareForExecutionImpl，创建一个新的 CodeBlock。

```
CodeBlock* codeBlock = newCodeBlockFor(kind, function, scope, exception);
resultCodeBlock = codeBlock;
EXCEPTION_ASSERT(!!throwScope.exception() == !codeBlock);
if (UNLIKELY(!codeBlock))
   return exception;
```

```
if (Options::validateBytecode())
    codeBlock->validate();

if (Options::useLLInt())
    setupLLInt(vm, codeBlock);
else
    setupJIT(vm, codeBlock);

installCode(vm, codeBlock, codeBlock->codeType(),
codeBlock->specializationKind());
```

至此，如果 Option 指定为 LLInt，那么就会调用 setupLLInt 方法设置 CodeBlock，在其他情况下，则会直接通过 JIT 编译。LLInt 在 LLIntSlowPaths.cpp 里通过 C 语言函数封装了 LowLevelInterpreter.asm 里的执行指令的汇编代码。触发 JIT 优化是通过热点探测方法实现的。LLInt 会在字节码的 loop_hint 循环计数和 ret 函数返回指令的时候进行统计，结果保存在 ExecutionCounter 里。当函数或循环体执行到一定次数时，通过 checkIfThresholdCrossedAndSet 方法得到布尔值的结果来决定是否用 JIT 编译。

CodeBlock 有 GlobalCode、EvalCode、FunctionCode、ModuleCode 几种类型，用于 LLInt 和 JIT。编译后的字节码会保存到 UnlinkedCodeBlock 里。

3.3.6 分层编译

总的来说 JavaScriptCore 采用了类型推测和分层编译的思想，解析成字节码后 LLInt 的作用就是让 JavaScript 代码能够早点执行。由于解释的效率不高，所以达到一定条件后可以并行地通过 Baseline JIT 编译成更高效的字节码来解释，如果出现使 LLInt 解释效率差的情况，就会并行使用 DFG JIT。DFG 编译的结果也有代表 On-stack replacement 的 osrExitSite，这样的字节码的解释性能会更好，假如解释失败还可以再回到 Baseline。启用 FTL 的条件就更高了，可能函数调用好几万次都不会开启。

分层编译的主要思想是先把源代码解释成一种内部的表示，也就是字节码，将其中使用率高的字节码转成汇编代码，汇编代码可以由 CPU 直接执行以提高性能。

LLInt、Baseline JIT 和 DFG JIT 三者会同时运行。可以在 jsc.cpp 里添加日志观察它们的工作状态。

箭头指示的是堆栈替换（On-Stack Replacement），简称 OSR，如下图所示。这个技术可以将执行转移到任何 statement 的地方。OSR 可以不管执行引擎是什么，都处于解析字节码状态，并且能够重构它，让其他引擎继续执行。OSR entry 是进入更高层优化，OSR exit 是降至低层。从 LLInt 转到 Baseline JIT 时，OSR 只需要跳转到相应的机器代码地址即可，因为 Baseline JIT 每个指令边界上的所有变量的表示方式和 LLInt 是一样的。进入 DFG JIT 以后流程就会更加复杂，通过函数控制流图可以发现多个入口，一个是函数启动入口，一个是循环的入口。

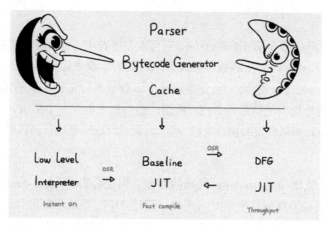

如果每个函数都用 DFG JIT 来优化，性能开销太大，DFG JIT 的优化过程性能开销和本地应用源代码一次完整编译的性能开销一样。如果每个语句都只执行一次，那么 DFG JIT 编译所需的时间就会比 LLInt 需要的时间长。执行的次数一多，LLInt 这种解释方式就会比编译方式的效率差，LLInt 的大部分时间都消耗在分派下个字节码指令的操作上了。

DFG JIT

DFG JIT 是一种推测优化的技术。DFG JIT 会开始对一个类型进行假设，类型假设的原则是优先选择利于性能优化的类型。按照利于性能优化的类型，先编译一个版本，如果后面发现假设不对就会跳转回原先的代码，这一过程称为 "Speculation Failure"。DFG 是并发编译器，DFG pipeline 的每个部分都是同时运行的，包括字节码解析，以及对字节码解析后结果的分析。DFG JIT 的优化方式在开始时会把字节码设置成 DFG CPS 格式，这个格式的字节码会描述变量和临时数据之间的数据流关系。再用分析信息来推测类型，通过推测的类型减少类型的检查。接下来进行的是传统的编译器方面的优化工作，最后编译器通过 DFG CPS 格式的字节码直接生成机器码。

DFG JIT 会将字节码解析成 SSA 形式的文件，按执行路径收集类型信息。DFG JIT 还会使用 inlining 和 value profiling 等一些静态分析技术。类型推导把 value profile 里常用的类型假设成后面将要使用的类型，使用常用类型进行预测，在 SpeculatedType.h 里定义一些数据类型，并使其符合 Data-flow analysis 规范，具体实现在 DFGPredictionPropagationPhase.cpp 里。Baseline JIT 次数超过一定数量就会生成一个新的类型，可能会触发 DFG，也可能会让 Baseline JIT 再执行几次。

DFG 的 register allocator 中汇编程序使用的是 op src1、src2 这样的二地址形式，不用 op dest、src1、src2 这样的三地址形式，结果存储在第一个操作数里。

DFG JIT 将低效字节码转成更高效的形式。DFG 结合编译器的优化方式，比如寄存器的分配、控制流图的简化、公共子表达式消除、死代码消除和稀疏条件常量传播等。但是正规的编译器是需要知道变量类型和堆里面的对象的结构的，这样才能更好地优化。DFG JIT 使用的是 LLInt 和 Baseline JIT 里分析出的变量类型进行推断的。比如下面的例子：

```
function plusThe(x) {
    return x + 1;
```

}

这里的 x 从代码上看是看不出类型的，它可能是字符串，也可能是整数型或者浮点型变量。LLInt 和 Baseline JIT 会从函数参数或者来自堆的读取来收集每个变量的值信息。DFG 不会一开始就编译 plusThe 函数，它会等到这个函数被执行了很多次后编译，让 LLInt 和 Baseline JIT 尽可能地把类型信息收集全。根据这些信息来假设一个类型，一个确定的类型会节省空间和执行时间。如果假设的检查失败，那么就通过 OSR 把执行转移到 Baseline JIT 上。

FTL JIT

DFG JIT 可以优化 long-running 代码的性能，但是会影响 short-running 代码的性能，DFG 生成代码的质量不高而且生成时间还长。想让 DFG 生成更好的代码会降低代码速度，还会增加 shorter-running 代码的延迟。DFG 不可能同时成为低延迟优化和高流通量的代码编译器。这时就需要再多一层 JIT 来做"较重"的优化，同时让 DFG 保持自己的"体重比较轻"，使得 long-running 和 short-running 代码的性能能够保持平衡。

新的一层 FTL 实际上是 DFG Backend 的替换。在 DFG 的 JavaScript 函数转换为静态单一指派（SSA）格式时，JavaScriptCore 会做 JavaScript 特性优化。接着 JavaScriptCore 把 DFG IR 转换成 FTL 里用到的 B3 的 IR，最后生成机器码。

DFG 的机器码生成器运行很快，但没有进行低级优化。对机器码的优化采用转成 SSA 的方式，比如自动执行循环不变量代码来提高执行速率，即使用被称为循环不变量代码移动的优化技术。优化完成后就可以进行 straight-forward linear 了，并且可以一对多（DFG SSA 的每个指令都可以产生）地转换为同样基于 SSA 却没有 JavaScript 语言特性的 B3 的 IR。总的来说，整个过程就是把源代码生成字节码，生成的字节码会转换成 DFG CPS IR。DFG CPS IR 下一步是转换成 DFG SSA IR，最终转换成 B3 的 IR。JavaScript 的动态性就是在这一过程中一步步被消除掉的，如下图所示。

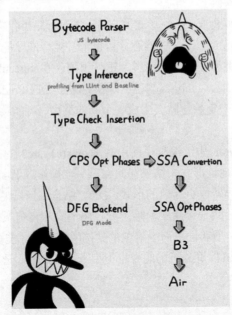

代码在 LLInt、Baseline JIT 和 DFG JIT 运行一段时间才会调用 FTL。FTL 在并发线程中会在 LLInt、Baseline JIT、DFG JIT 运行期间收集、分析信息。short-running 代码是不会导致 FTL 编译的，一般超过 10 毫秒的函数会触发 FTL 编译，如下图所示。

FTL 通过并发编译，减小对启动速度的影响。

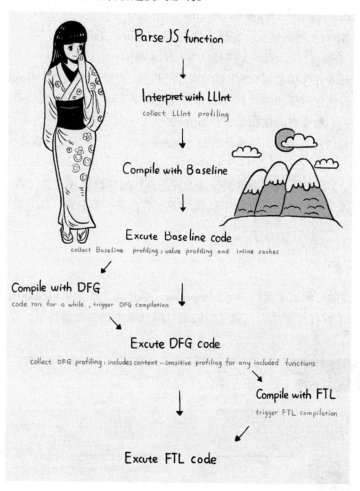

FTL Backend B3

B3 的全称为 Bare Bones Backend，实现了大吞吐量同时缩短总体 FTL 编译时间。B3 的 IR 是一个低级别的中间表示，低级别意味着需要大量的内存表示每个函数。需要大量的内存意味着编译器在分析函数时需要扫描大量内存。B3 有两个 IR，一个代码级别的 IR 叫 B3 IR，另一个机器级别的 IR 叫 Assembly IR，简称 Air。这两个 IR 的目标是实现低级操作，同时能最大限度地减少分析代码所需要的内存。为达到这个目标，需要减少 IR 对象的整体大小，减少表示典型操作的 IR 对象数量，减少 IR 的指针跟踪，最后减少 IR 的总数。

B3 转换成 Air 会通过执行反向贪婪模式匹配，巧妙地针对每个 B3 操作选择正确的 Air 序列。具体实现可以参考 B3LowerToAir.cpp 的代码。

下面我们来看看 B3 IR 所做的优化。

- Strength reduction：主要包括控制流程图的简化，常量合并，消除死代码，消除整数溢出检查，还有其他简化规则。实现文件是 B3ReduceStrength.cpp。
- Flow-sensitive：流敏感，会对常量进行合并，实现在 B3FoldPathConstants.cpp 文件里。
- Eliminate Common Subexpressions：全局共用子表达式消除，实现文件是 B3EliminateCommonSubexpressions.cpp。
- Tail duplication：尾部重复，实现文件是 B3DuplicateTails.cpp。
- Fix SSA：SSA 修正，实现文件是 B3FixSSA.cpp。
- Move constants：移动常量，优化常量实际的位置，实现文件是 B3MoveConstants.cpp。
- Air 里做的优化包括去除死代码、简化控制流图，以及修正部分寄存器和减轻溢出代码症状。Air 最重要的优化是寄存器分配。

B3 IR 的动态语言支持，一半是靠的是 Patchpoint，另一半是靠的是 OSR，OSR 所做的就是用操作码的 Check family 实现栈上的替换。JavaScript 是种动态语言，没有可以快速执行的类型和操作，FTL JIT 就使用推测行为方式把这些操作转换成快速代码。

Air 在寄存器分配的选择上使用了经典的图形着色寄存器分配，即迭代寄存器合并，简称 IRC。

3.3.7　类型分析

WebKit 里提供的调试工具叫 Web Inspector，通过这个工具可以很好地观察哪些函数甚至哪些条件执行了或者没有执行。最重要的是可以直观地观察变量类型，以及跟踪值的继承链，如下图所示。

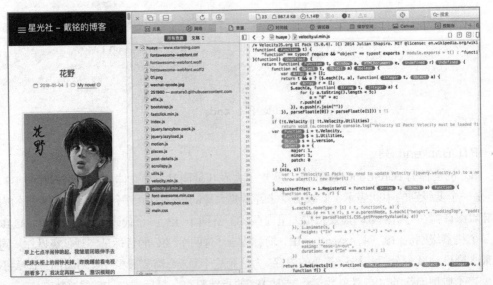

JavaScript 是动态类型语言，变量可以是任何类型，表达式可以有任何类型，函数可以返回任何类型，等等。由于不能立刻得到类型，没法确定地址字节大小，所以需要在运行时进行反复推断。静态语言就不一样了，一开始就明确了类型，比如目标平台中的 int 类型占用

4 个字节，它对应的对象是个地址，偏移量是 0，访问 int 时需要将对象地址加上 4 个字节。所以语言解释系统只要用数组和位移来存储变量和方法地址就行了，这样几个机器语言的指令就可以对其执行各种操作了。

我们看下面的例子。

```
let x = 3;
print(x);
x = ["These", "is", "array with string"];
print(x);

const justReturn = (x) => { return x; };
justReturn(10);
justReturn([1, 2, 3]);
justReturn("I am bad!");
```

JavaScript 没有类型限制，语法不出错就行。如果是静态类型语言，就没法像上面的例子一样把数组赋给数字类型的变量。在 Type Profiler 中会高亮显示类型名为"String?"和"Integer?"，像这样后面跟着问号的类型。这是在有多个不同类型被分配给相同变量、函数参数传递和函数返回时发生的。在查看时，可以看到所有这些被收集的类型信息。

JavaScriptCore 是如何分析变量类型的呢？我们先看一段 JavaScript 代码。

```
function add(x, y) {
    return x + y;
}
```

对应的字节码如下所示。

```
function add: 10 m_instructions; 3 parameter(s); 1 variable(s)
[ 0] enter
[ 1] get_scope        loc0
[ 3] add              loc1, arg1, arg2
[ 8] ret              loc1
```

开启了 Type Profiler 后的字节码如下所示。

```
function add: 40 m_instructions; 3 parameter(s); 1 variable(s)
[ 0] enter
[ 1] get_scope        loc0

// 分析函数参数
[ 3] op_profile_type  arg1
[ 9] op_profile_type  arg2

// 分析 add 表达式的操作数
[15] op_profile_type  arg1
[21] op_profile_type  arg2
[27] add              loc1, arg1, arg2

// 分析返回语句，收集这个函数的返回类型信息
[32] op_profile_type  loc1
[38] ret              loc1
```

进入 DFG 时转换成 DFG IR 的代码如下所示。

```
 0: SetArgument(this)
 1: SetArgument(arg1)
 2: SetArgument(arg2)
 3: JSConstant(JS|PureInt, Undefined)
 4: MovHint(@3, loc0)
 5: SetLocal(@3, loc0)
 6: JSConstant(JS|PureInt, Weak:Cell: 0x10f458ca0 Function)
 7: JSConstant(JS|PureInt, Weak:Cell: 0x10f443800 GlobalScopeObject)
 8: MovHint(@7, loc0)
 9: SetLocal(@7, loc0)
10: GetLocal(JS|MustGen|PureInt, arg1)
11: ProfileType(@10)
12: GetLocal(JS|MustGen|PureInt, arg2)
13: ProfileType(@12)
14: ProfileType(@10)
15: ProfileType(@12)
16: ValueAdd(@10, @12, JS|MustGen|PureInt)
17: MovHint(@16, loc1)
18: SetLocal(@16, loc1)
19: ProfileType(@16)
20: Return(@16)
```

现在还有很多 ProfileType，这些操作会有很大的消耗，DFG 会推测参数类型是整数。接下来，在这个假设下将 ProfileType 的操作移除。

```
 1: SetArgument(arg1)
 2: SetArgument(arg2)
 3: JSConstant(JS|PureInt, Undefined)
 4: MovHint(@3, loc0)
 7: JSConstant(JS|PureInt, Weak:Cell: 0x10f443800 GlobalScopeObject)
 8: MovHint(@7, loc0)
10: GetLocal(@1, arg1)
12: GetLocal(@2, arg2)
16: ArithAdd(Int32:@10, Int32:@12)
17: MovHint(@16, loc1)
20: Return(@16)
```

在对类型的处理上，V8 用一个通过字符串匹配属性算法的隐藏类，结合 C++ 类和偏移位置思想对类型进行处理。具体做法是将有相同属性名和属性值的对象保存在同一个组的隐藏类里，这些属性在隐藏类里有着同样的偏移值。这个组里的对象能够共用这个隐藏类的信息。访问属性的过程是得到隐藏类的地址，根据属性名得到偏移值，通过偏移值和隐藏类地址得到属性地址。这个过程是否可以加速呢？答案是肯定的，通过 Inline Cache 缓存之前的查找结果可以减少函数和属性的哈希表查找时间。当一个对象或属性出现多种类型时，缓存会不断更新。当发生这种情况时，V8 就会按照先前哈希表查找的方式来查找。

3.3.8 指令集架构

JavaScriptCore 是基于寄存器的虚拟机 register-based VM。这种实现方式不用频繁入栈、出栈和执行三地址的指令集，所以效率高，但移植性弱一点。

基于寄存器指令集架构

三地址和二地址的指令集基本都是使用基于寄存器的架构来实现的。要求是除 load 和 store 以外的运算指令的源都要是寄存器。

代码如下所示。

```
i = a + b;
```

将上面这句代码转换成机器指令，得到的结果如下所示。

```
add i, a, b
```

这样的形式就是三地址指令，很多代码都是这样的二元运算，然后再进行赋值。三地址正好可以分配两地址给二元运算的两个源，剩下一个地址给赋值目标。ARM 处理器的主要指令集就是三地址的。那么二地址是怎么处理的呢？将上面的代码换成下面的样子。

```
i += a;
i += b;
```

机器指令就变成了以下形式。

```
add i, a
add i, b
```

上面代码中的 i 同时也作为赋值目标。x86 系列的处理器就是采用的二地址。

有了二地址和三地址，有一地址吗？下面就是一地址。

```
add a
add b
```

只有操作源，那么目标在哪儿呢？目标 i 是隐藏目标，这种运算的目标被称为"累加器的专用寄存器"，运算依赖于更新累加器的状态来完成。

基于栈指令集架构

有零地址吗？JVM 采用的就是零地址。我们先看一段 Java 代码。

```
class QuickCalculate {
  byte onePlusOne() {
    byte x = 1;
    byte y = 1;
    byte z = (byte) (x + y);
    return z;
  }
}
```

转换成字节码，如下所示。

```
iconst_1    // Push 整数常量1
istore_1    // Pop 到局部变量1, 相当于 byte x = 1; iconst_1
            // 再 Push 整数常量1
istore_2    // Pop 到局部变量2, 相当于 byte y = 1;
iload_1     // Push x, 这时 x 已经作为整数存在局部变量1里
```

```
iload_2     // Push y, 这时 y 已经作为整数存在局部变量 1 里
iadd        // 执行加操作, 栈顶就变成了 x + y, 结果为一个整数
int2byte    // 将整数转化成字节, 结果还是占 32 位
istore_3    // Pop 到局部变量 3 里, byte z = (byte)(x + y)
iload_3     // Push z 的值使之可以被返回
ireturn     // 返回结果, return z
```

可以看到零地址形式的指令集是基于栈架构的。这种架构的优势是可以用更少的空间存储更多的指令,所以在空间不是很充足时这种架构是可取的,不过零地址要完成一件事会比基于寄存器指令架构的二地址、三地址指令要多很多指令,执行效率更低。

指令集架构演化和比较

JavaScriptCore 采用 SquirrleFish 之前的解释器, 它们都是树遍历式的, 解释器会递归遍历树。遍历树上的每个节点, 所做的操作就是解释每个节点返回的值。这种操作既不是基于栈, 也不是基于寄存器的。例如:

```
i = x + y * z
```

按 AST 的后序遍历, 最上面的 = 符号节点返回值依赖 + 符号节点返回的值, + 符号依赖 x 节点和 * 符号节点的值, 依此递归下去。这样最开始获取到值的运算就是最低一级的运算。在这个例子里就是将 y 和 z 的运算结果返回给 * 符号节点。这就是典型的后序遍历。Ruby 1.9 之前也是用这种方式解释执行的。

赋值符号=左侧的值称为"左值", 赋值符号=右侧的值称为"右值"。左值也可能是复杂的表达式, 比如数组或者结构体。根据求值顺序, 对于二元运算的节点, 是先遍历左子节点的。所以当左值是复杂表达式需要计算时是会优先进行计算的。我们再看一个左值是数组的例子, 代码如下所示。

```
public class LeftFirstTest {
    public static void main(String[] args) {
        int[] arr = new int[1];
        int x = 3;
        int y = 5;
        arr[0] = x + y;
    }
}
```

arr[0] = x + y; 对应的字节码如下所示。

```
// 左值, 数组下标
aload_1
iconst_0

// 右值
iload_2 // x
iload_3 // y
iadd

iastore   // 赋值符号节点进行赋值
```

可以看到左值是先计算的。

从树遍历到基于栈的解释，实际上是将 AST 的树结构转换成了平行的栈结构。方法是在后序遍历 AST 时使用 Reverse Polish Notation。这种后缀记法生成一个序列，成为一个线性结构，之后再解释执行这个操作序列。

JVM 是 Java 语言的虚拟机。开发 Android 应用程序使用的是 Java 语言，那么它的虚拟机也是 JVM 吗？答案是否定的，Android 使用的虚拟机叫 Dalvik VM，这款虚拟机在很多设计上都与 JVM 兼容，字节码是基于寄存器架构的，既可以是二地址指令，也可以是三地址指令。Dalvik VM 用于移动端，为了能够更加高效，所以从开始时就没有考虑太多可移植性，因此基于寄存器架构的优势就能够更好地发挥出来了。如果想了解更多关于 Dalvik VM 的知识，可以看 Dan Bornstein 做的关于 Dalvik 实现原理的演讲"Dalvik VM Internals"。

JVM 和 Dalvik VM 的主要区别是后者字节码指令数量更少、内存更小。JVM 每个线程都有一个 Java 栈，叫 activation record。activation record 是专门用来记录方法调用的，每调用一个方法就会分配一个新栈帧，方法返回就 Pop 出栈帧。每个栈帧会有局部变量区，istore 指令的作用是将局部变量和参数移到局部变量区。每个栈帧还会有求值栈，这个栈用来存储求值的中间结果和调用其他方法的参数等。使用 iconst 指令来进行数据移动，还可以通过 iadd、imul 指令在求值栈中 Pop 出值进行求值，然后再把结果 Push 到栈里。

Dalvik VM 的每个线程有程序计数器和调用栈，方法的调用和 JVM 一样会分配一个新的帧，不同的是 Dalvik VM 使用的是虚拟寄存器来替代 JVM 里的局部变量区和求值栈。方法调用会有一组自己的虚拟寄存器，常用的是 v0～v15，有些指令可以使用 v0～v255。只在虚拟寄存器中进行指令操作，数据移动少了，保存局部变量的存储单元也会少很多。Dalvik VM 在每次方法调用时都会调用一组自己的寄存器。不过在 x86 架构中寄存器是全局描述符表，因此 x86 需要考虑 calling converntion，即需要在调用时保护一些寄存器的状态。而 Dalvik VM 在每次调用方法后，那些寄存器值会恢复成调用前的状态，避免发生错误。

V8 没有中间的字节码，而是直接编译 JavaScript 生成机器码。不过在内部也用了表达式栈来简化代码生成，在编译过程中使用虚拟栈帧来记录局部变量和栈的状态。生成代码的过程中会有窥孔优化来去除多余的入栈和出栈，把栈操作转成寄存器操作，生成的代码就和基于寄存器的代码类似了。

至此，会有人产生疑问，零地址要做多次入栈和出栈的操作，执行效率低，为什么还会有虚拟机（比如 JVM）要用零地址形式呢？

这是因为 x86 之类的处理器最初的寄存器很多不是通用寄存器，需要让编译器决定程序里那么多的变量该怎么装到寄存器里，哪些应该映射到一个寄存器，哪些应该换出，要解决这些问题并不容易，JVM 采用零指令这种堆栈结构的原因就是不信任编译器的寄存器分配方案。使用堆栈结构，就可以避开寄存器分配的难题。不过后来 IBM 公开了他们的图染色寄存器分配算法，才使编译器的分配能力得到很大的提升，所以现在都是编译器来主导寄存器分配的。

很多主流高级语言虚拟机（比如 JVM、CPython）采用的是基于栈的架构。主要是因为这样不用考虑给临时变量分配空间，只需要求值栈来完成指令即可，这样的编译器更容易实现，并且更容易在硬件较差的机器上运行，前面讲到基于栈的架构指令对于存储空间的要求更低。最后考虑的是移植性，复杂指令集计算机（Complex Instruction Set Computer，简称 CISC）通用寄存器的数量比较少，32 位复杂指令集计算机只有 8 个 32 位通用寄存器，而精简指令集计算机（Reduced Instruction Set Computer，简称 RISC）通用寄存器的数量会多一些，32 位精简指令集计算机有 16 个寄存器。源架构寄存器的数量通常和实际机器通用寄存器的数量不一致且映射实现起来很麻烦。由于栈架构里是没有通用寄存器的，所以实现虚拟机时很容易自由地分配实际机器寄存器，移植起来自然也就容易很多。

大多数处理器是基于寄存器架构的。但是对于虚拟机来说，基于栈架构需要执行更多的 load 或 store 指令，这样指令分配的次数和内存访问的次数会更多。基于寄存器的架构在虚拟机里更好些。苹果公司的产品都在自己的闭环里，苹果公司并不太在意技术的移植性，所以 JavaScriptCore 会选择性能更好的基于寄存器的架构，而不会为了架构移植性而牺牲性能。

3.3.9　JavaScript

各个 JavaScript 引擎的介绍

- SpiderMonkey：用于 Mozilla Firefox，是最早的 JavaScript 引擎，SpiderMonkey 使用的字节码是基于栈的架构的。Parser 使用的是手写纯递归下降式。
- KJS：KDE 的引擎，用于 Konqueror 浏览器。树遍历解释器。Parser 使用的是 bison。
- Rhino：使用 Java 编写，开放源代码，也是 Mozilla 的 JavaScript 引擎，相当于 Java 版的 SpiderMonkey。Mozilla 曾将 JavaScript 作为服务端的脚本语言来用。Parser 使用的是手写的纯递归下降式。
- Chakra：也叫 JScript，微软的 Internet Explorer 和 Microsoft Edge 都在使用。
- JavaScriptCore：苹果公司开发的开源 JavaScript 引擎，用在 Safari 等浏览器中。Safari 可以开启 JavaScriptCore 的 JIT，而 UIWebView 则无法开启 JavaScriptCore 的 JIT。
- V8：Google 开发的开源引擎，用于 Chrome。Parser 使用的是手写纯递归下降加运算符优先级混合式。V8 使用了 2-pass，会先收集一些上下文信息以增加预测的准确性。V8 最初直接将代码编译成机器码，Dart VM 也是这样设计的。不过 V8 5.9 启用了 Ignition 字节码解释器。自此，几大 JavaScript 引擎都用了字节码。启用字节码主要是希望减少机器码对内存空间的占用。由于机器码占用的空间很大，所以不好都缓存下来，不然内存和磁盘都吃不消，序列化和反序列化时间都太长，每次编译的机器码都是不完整的，只会缓存最外的一层。如果代码最外层包了一层，启动代码每次都是不同的调用就会每次都编译，导致缓存没有起作用。由于字节码经过精心设计后比机器码更紧凑，所以在引入 Ignition 后内存占用的空间明显下降了。此时 TurboFan 再对 Ignition 字节码进行解释。

各个引擎之间的通用技术

根据下面的关键字查找资料，可以了解更多的虚拟机技术。

- Recusive-descent Parser：递归下降式 Parser。
- Operator precedence Parser：运算符优先级 Parser。

- Deferred Parser：延迟 Parser。
- Tiered Compilation：多层编译。
- Background Compilation：后台编译。
- Type Feedback：类型反馈。
- Type Specialization：类型特化。
- SSA-form IR：静态单赋值形式。

3.4　WebCore

3.4.1　浏览器历史

1990 年，Berners-Lee 发明了 WorldWideWeb 浏览器，后改名为"Nexus"并在 1991 年公布了源代码。

1993 年，Marc Andreessen 的团队开发了 Mosaic，1994 年推出了 Netscape Navigator 浏览器，其在最火爆的时候曾是绝大多数人首选的浏览器。

1995 年，微软推出了 Internet Explorer 浏览器，简称 IE。通过免费绑定进 Windows 95 系统替代 Netscape，引发了浏览器大战。

1998 年，网景公司成立了 Mozilla 基金会组织，同时开源了浏览器代码，2004 年推出有多标签并支持扩展的 Firefox 浏览器，开始慢慢占领市场。

2003 年，苹果公司发布了 Safari 浏览器，2005 年放出了核心源代码。

2008 年，Google 以苹果公司的 WebKit 为内核，建立了新的项目 Chromium，在此基础上开发了自己的浏览器 Chrome。

2012 年，WHATWG 和 W3C 推出 HTML5 技术规范。HTML5 包含了 10 大特性：离线、存储、链接、文件访问、语义、3D、图形、展示、性能、基本要素。

3.4.2　WebKit 全貌

Webkit 架构

从浏览器技术的发展可以看到，到了后期，浏览器基本都是基于 WebKit 开发的，那么我们就来了解一下 WebKit。它的完整架构如下图所示。

WebCore 是共享的,其他部分会根据不同的平台有不同的实现。
下图为 WebKit 最核心、最复杂的排版引擎 WebCore 的结构图。

解析 HTML 会产生 DOM Tree,解析 CSS 会产生 CSS Rule Tree。JavaScript 会通过 DOM API 和 CSS Object Model(CSSOM) API 来操作 DOM Tree 和 CSS Rule Tree。

Layout 主要包含定位坐标和视图大小,比如 position、overflow、z-index 等。动态修改

DOM 属性和 CSS 属性会导致重新布局。

WebKit 源代码结构说明

- JavaScriptCore：默认 JavaScript 引擎，Google 已经使用了 V8 作为其 Chromium 的 JavaScript 引擎。
- WebCore：浏览器渲染引擎，包含了各个核心模块。
- WebCore/css：包括 CSS 解释器、CSS 规则等。
- WebCore/dom：各种 DOM 元素和 DOM Tree 结构相关的类。
- WebCore/html：HTML 解释器和各种 HTML 元素等相关内容。
- WebCore/rendering：Render Object 相关的内容，还有页面渲染的样式和布局等。
- WebCore/inspector：网页调试工具。
- WebCore/loader：指主资源和派生资源的加载，以及派生资源的 MemoryCache 等。
- WebCore/page：页面相关的操作、页面结构和交互事件等。
- WebCore/platform：各个平台相关的代码，比如 iOS 系统相关代码和 macOS 系统相关代码等。
- WebCore/storage：与存储相关的功能，比如 WebStorage、Index DB 等接口的实现。
- WebCore/workers：Worker 线程封装，提供 JavaScript 多线程执行环境。
- WebCore/xml：XML 相关的功能，比如 XML Parser、Xpath、XSLT 等。
- WebCore/accessibility：图形控件访问接口。
- WebCore/bindings：DOM 元素和 JavaScript 绑定的接口。
- WebCore/bridge：C 语言、JavaScript 和 Objective-C 的桥接。
- WebCore/editing：与页面编辑相关的功能，比如修改 DOM、拼写检查等。
- WebCore/history：利用 Page Cache 实现跳转、浏览历史记录等。
- WebCore/mathml：数学表达式在网页中的代码实现。
- WebCore/plugins：NPPlugin 的支持接口。
- WebCore/svg：对矢量图形的支持。
- WebKit：平台相关的接口，每个目录都是不同平台的接口实现。
- WTF：基础类库，类似 C++ 的 STL 库，比如字符串操作、智能指针、线程等。
- DumpRenderTree：用于生成 RenderTree。
- TestWebKitAPI：测试 WebKit 的 API 的测试代码。

WebKit 代码风格

- 内存管理：使用的引用计数技术，在 RefCounted 模板类里有 ref() 加和 unref() 减来进行控制，很多类都是继承这个模板类的。后期加入了 RefPtr 和 PassRefPtr 智能指针模板类，采用重载赋值操作符的方式自动完成计数加减，这样基本没有发生内存泄漏的可能。
- 代码自动生成：C++ 对象到 JavaScript 对象的 Binding 实现，使用的是代码自动生成。依据 IDL 接口描述文件对接口的描述，用 Pearl 脚本完成代码的自动生成。部分 CSS

和 HTML 解析也用到了代码自动生成，这样添加 CSS 属性值和 HTML 标签属性能节省大量时间，只需要修改 .in 配置文件即可。
- 代码编写风格：可以在 WebKit 官网查到，想给 WebKit 做贡献的人可以参考。

WebKit 设计模式

下面我们看一看在 WebKit 里使用了哪些设计模式，以及它们是如何使用的。

- 单例模式：WebKit 里的 Loader 管理 CacheResource 就是单例模式。
- 工厂模式：可以在 WebKit 源代码里搜索结尾是 Factory 的代码，一般用的是工厂模式。
- 观察者模式：代码名称的结尾是 Client 的都是观察者模式，比如 FrameLoaderClient 可以看成是观察者类，被观察者 FrameLoader 会向 Client 观察者类通知自身状态的变化。
- 组合模式：用于树状结构对象，比如组成 DOM Tree 和 Render Tree 的类对象分别是 ContainerNode 和 RenderObject，ContainerNode 和 RenderObject 之间的联系使用的就是组合模式。
- 命令模式：DOM 模块的 Event 类和 Editing 模块的 command 类都是命令模式。

WebKit 主要类

WebKit 的主要类如下图所示。

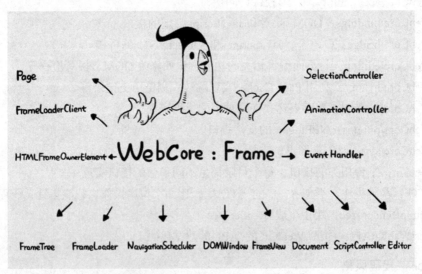

- Frame：中心类，通过它找其他类。
- FrameLoader：加载资源用的。
- FrameView：处理绘制和滚动等。
- Document：具体实现是 HTMLDocument。
- Page：窗口的操作，包括前进、后退、拖动等。
- EventHandler：处理输入事件，比如键盘、鼠标、触摸屏等。

- FrameTree：管理父 Frame 和子 Frame 的关系，比如 main frame 里的 iframe。
- FrameLoader：Frame 的加载。
- NavigationScheduler：主要用来管理页面跳转，比如重定向、meta refresh 等。
- DOMWindow：管理 DOM 相关的事件、属性和消息。
- FrameView：Frame 的排版。
- Document：用来管理 DOM 里的 Node，每个 tag 都会有对应的 DOM Node 关联。
- ScriptController：管理 JavaScript 脚本。
- Editor：管理页面，比如复制、粘贴和输入等编辑操作。
- SelectionController：管理 Frame 里的选择操作。
- AnimationController：动画控制。
- EventHandler：处理事件，比如鼠标、按键、滚动和缩放等事件。

下图展示了渲染的过程。

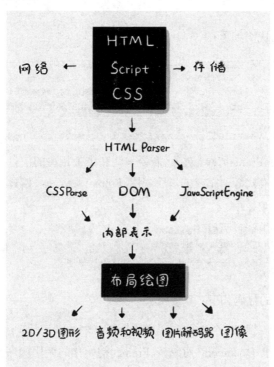

Frame 的主要接口

Create 接口的代码如下：

```
Ref<Frame> Frame::create(Page* page, HTMLFrameOwnerElement* ownerElement, FrameLoaderClient* client)
{
    ASSERT(page);
    ASSERT(client);
    return adoptRef(*new Frame(*page, ownerElement, *client));
}
```

在 Frame::create 里会调用 Frame 的构造函数，创建 Frame 对象。初始调用顺序如下：

```
webPage::setView
webPage::setViewportSize
webPage::mainFrame
webPagePrivate::createMainFrame
webFrameData::webFrameData
Frame::create
```

解析中发现 iframe 时的调用顺序如下：

```
FrameLoader::finishedLoading
HTMLDocumentParser::append
HTMLTreeBuilder::processToken
HTMLElementBase::openURL
SubFrameLoader::requestFrame
FrameLoaderClient::creatFrame
webFrameData::webFrameData
Frame::create
```

createView 接口的代码如下：

```
void Frame::createView(const IntSize& viewportSize, const Color&
  backgroundColor, bool transparent,
    const IntSize& fixedLayoutSize, const IntRect& fixedVisibleContentRect,
    bool useFixedLayout, ScrollbarMode horizontalScrollbarMode, bool horizontalLock,
    ScrollbarMode verticalScrollbarMode, bool verticalLock)
```

创建出用于排版的 FrameView 对象。需要一些和排版相关的信息，比如初始 viewport 的大小、背景色、滚动条模式等，完成创建后调用 Frame::setView 设置为当前 FrameView。调用顺序如下：

```
FrameLoader::commitProvisionalLoad
FrameLoader::transitionToCommitted
FrameLoaderClient::transitionToCommittedForNewPage
Frame::createView
```

setDocument 接口的代码如下：

```
void Frame::setDocument(RefPtr<Document>&& newDocument)
```

用来关联 Frame 和 Document 对象，Frame 初始化的调用顺序如下：

```
WebFrame::webFrame
webFramePrivate::init
Frame::init
FrameLoader::init
DocumentWriter::begin
Frame::setDocument
```

JavaScript 脚本更改数据时的调用顺序如下：

```
DocumentLoader::receivedData
DocumentLoader::commitLoad
FrameLoaderClient::committedLoad
```

```
DocumentLoader::commitData
DocumentWriter::setEncoding
DocumentWriter::willSetEncoding
FrameLoader::receivedFirstData
DocumentWriter::begin
FrameLoader::clear
Frame::setDocument
```

3.4.3 WTF

WTF 的全称是 Web Template Library,是 WebKit 的基础库,其实现了智能指针、字符串和容器,提供跨平台原子操作,以及时间和线程的封装,能够高效地对内存进行管理。WebCore 里都是 WTF 的代码而没有 STL 的代码。

智能指针

利用对原生指针的封装,C++ 可以自动化地实现资源管理和指针衍生操作,比如拷贝、引用计数、自动锁和延迟计算等。

智能指针主要作用是动态分配对象的内存并且自动回收,原理是智能指针的对象会作为栈上分配的自动变量在退出作用域时被自动析构。

智能指针的实现方式如下:

```
template <typename T>
class SmartPtr {
    public:
        typedef T ValueType;
        typedef ValueType *PtrType;

        //构造析构函数
        SmartPtr() :m_ptr(NULL) {}
        SmartPtr(PtrType ptr) :m_ptr(ptr) {}
        ~SmartPtr() {
            if(m_ptr) delete m_ptr;
        }

        //拷贝构造函数
        SmartPtr(const SmartPtr<T>& o); //堆上分配的对象
        template<typename U> SmartPtr(const SmartPtr<U>& o);

        //拷贝赋值运算符
        template<typename U> SmartPtr& operator = (const SmartPtr<U>& o);

        //指针运算,是为了让智能指针在行为上更类似原生指针
        ValueType& operator*() const {
            return *m_ptr;
        }
        PtrType operator->() const {
            return m_ptr;
        }

        //逻辑运算符重载
```

```cpp
        //对指针是否为空的判断 if(!ptr)
        //如果定义了 operator!(), 智能指针就可以用 if(!SmartPtr)
        bool operator!() const {
            return !m_ptr;
        }

        //转成 raw ptr
        operator PtrType() {
            return m_ptr;
        }
    private:
        PtrType m_ptr;
}
//创建智能指针对象格式
SmartPtr(new ValueType());
```

RefPtr

RefPtr 需要操作对象来引用计数，包含 ref() 和 deref() 方法的类对象才可以由 RefPtr 引用。为了不让 RefPtr 引用的类都用手工去添加 ref() 和 deref()，WTF 提供了 RefCounted 类模板，在 WTF/Source/wtf/RefCounted.h 里可以找到。类模板的定义如下：

```cpp
class RefCountedBase {
public:
    void ref() const
    {
#if CHECK_REF_COUNTED_LIFECYCLE
        ASSERT_WITH_SECURITY_IMPLICATION(!m_deletionHasBegun);
        ASSERT(!m_adoptionIsRequired);
#endif
        ++m_refCount;
    }

    bool hasOneRef() const
    {
#if CHECK_REF_COUNTED_LIFECYCLE
        ASSERT(!m_deletionHasBegun);
#endif
        return m_refCount == 1;
    }

    unsigned refCount() const
    {
        return m_refCount;
    }

    void relaxAdoptionRequirement()
    {
#if CHECK_REF_COUNTED_LIFECYCLE
        ASSERT_WITH_SECURITY_IMPLICATION(!m_deletionHasBegun);
        ASSERT(m_adoptionIsRequired);
        m_adoptionIsRequired = false;
#endif
    }
```

```cpp
protected:
    RefCountedBase()
        : m_refCount(1)
#if CHECK_REF_COUNTED_LIFECYCLE
        , m_deletionHasBegun(false)
        , m_adoptionIsRequired(true)
#endif
    {
    }

    ~RefCountedBase()
    {
#if CHECK_REF_COUNTED_LIFECYCLE
        ASSERT(m_deletionHasBegun);
        ASSERT(!m_adoptionIsRequired);
#endif
    }

    // Returns whether the pointer should be freed or not.
    bool derefBase() const
    {
#if CHECK_REF_COUNTED_LIFECYCLE
        ASSERT_WITH_SECURITY_IMPLICATION(!m_deletionHasBegun);
        ASSERT(!m_adoptionIsRequired);
#endif

        ASSERT(m_refCount);
        unsigned tempRefCount = m_refCount - 1;
        if (!tempRefCount) {
#if CHECK_REF_COUNTED_LIFECYCLE
            m_deletionHasBegun = true;
#endif
            return true;
        }
        m_refCount = tempRefCount;
        return false;
    }

#if CHECK_REF_COUNTED_LIFECYCLE
    bool deletionHasBegun() const
    {
        return m_deletionHasBegun;
    }
#endif

private:

#if CHECK_REF_COUNTED_LIFECYCLE
    friend void adopted(RefCountedBase*);
#endif

    mutable unsigned m_refCount;
#if CHECK_REF_COUNTED_LIFECYCLE
```

```
    mutable bool m_deletionHasBegun;
    mutable bool m_adoptionIsRequired;
#endif
};

template<typename T> class RefCounted : public RefCountedBase {
    WTF_MAKE_NONCOPYABLE(RefCounted); WTF_MAKE_FAST_ALLOCATED;
public:
    void deref() const
    {
        if (derefBase())
            delete static_cast<const T*>(this);
    }

protected:
    RefCounted() { }
    ~RefCounted()
    {
    }
};
```

RefCounted 类会被大量继承，任何类如果想被 RefPtr 引用只需要继承 RefCounted 类即可。在 RefPtr.h 里可以看到 refIfNotNull 和 derefIfNotNull 定义了计数的增加和减少。

Assert 的实现和应用

Assert 在 WTF 里的定义代码如下所示。

```
#define ASSERT(assertion) do { \
    if (!(assertion)) { \
        //打印用
        WTFReportAssertionFailure(__FILE__, __LINE__, WTF_PRETTY_FUNCTION,
#assertion); \
        //重点
        CRASH(); \
    } \
} while (0)

#ifndef CRASH
#if defined(NDEBUG) && OS(DARWIN)
// Crash with a SIGTRAP i.e EXC_BREAKPOINT.
// We are not using __builtin_trap because it is only guaranteed to abort,
// but not necessarily
// trigger a SIGTRAP. Instead, we use inline asm to ensure that we trigger
// the SIGTRAP.
#define CRASH() do { \
    //直接 inline 汇编代码
    WTFBreakpointTrap(); \
    __builtin_unreachable(); \
} while (0)
#else
#define CRASH() WTFCrash()
#endif
#endif // !defined(CRASH)
```

```
//根据不同 CPU 定义不同的 WTFBreakpointTrap() 汇编宏
#if CPU(X86_64) || CPU(X86)
#define WTFBreakpointTrap() __asm__ volatile ("int3")
#elif CPU(ARM_THUMB2)
#define WTFBreakpointTrap() __asm__ volatile ("bkpt #0")
#elif CPU(ARM64)
#define WTFBreakpointTrap() __asm__ volatile ("brk #0")
#else
#define WTFBreakpointTrap() WTFCrash() // Not implemented.
#endif
```

内存管理

WTF 提供的内存管理和 STL 类似，都为容器和应用提供内存分配接口。先了解一下 WTF_MAKE_FAST_ALLOCATED 宏，代码如下所示。

```
#define WTF_MAKE_FAST_ALLOCATED \
public: \
    void* operator new(size_t, void* p) { return p; } \
    void* operator new[](size_t, void* p) { return p; } \
    \
    void* operator new(size_t size) \
    { \
        return ::WTF::fastMalloc(size); \
    } \
    \
    void operator delete(void* p) \
    { \
        ::WTF::fastFree(p); \
    } \
    \
    void* operator new[](size_t size) \
    { \
        return ::WTF::fastMalloc(size); \
    } \
    \
    void operator delete[](void* p) \
    { \
        ::WTF::fastFree(p); \
    } \
    void* operator new(size_t, NotNullTag, void* location) \
    { \
        ASSERT(location); \
        return location; \
    } \
private: \
typedef int __thisIsHereToForceASemicolonAfterThisMacro
```

这个宏定义了 operator new 和 operator delete。operator new 是直接调用 fastMalloc 完成内存分配的。下面我们看一看 fastMalloc 的实现，代码如下所示。

```
void* fastMalloc(size_t size)
{
```

```
    ASSERT_IS_WITHIN_LIMIT(size);
    return bmalloc::api::malloc(size);
}
```

可以看到 WTF 使用了 bmalloc 方法进行 malloc 内存分配。据苹果公司宣传，这个 bmalloc 的性能要远超 TCMalloc，和 STL 里在 malloc 之上实现一个内存池的思路完全不一样。关于 STL 内存管理的实现，《STL 源码剖析》一书的第 2 章有详细说明，建议大家阅读。

容器类

WTF 里有很多的容器，比如 BlockStack、Deque、Vector 和 HashTable。

WTF 里的 Vector 和 STL 里的 Vector 一样也是一种数据结构，相当于动态数组。当不知道需要的数组的规模时，Vector 可以达到节约空间的目的。我们先看一看 WTF 实现 Vector 的方法。关键地方笔者在代码里做了注释说明，代码如下所示。

```
template<typename T, size_t inlineCapacity = 0, typename OverflowHandler =
CrashOnOverflow, size_t minCapacity = 16, typename Malloc = FastMalloc>
    class Vector : private VectorBuffer<T, inlineCapacity, Malloc> {
        WTF_MAKE_FAST_ALLOCATED;
private:
//VectorBuffer 是内部存储数据的容器
    typedef VectorBuffer<T, inlineCapacity, Malloc> Base;
    //Vector 里元素的初始化、复制和移动等操作都在 VectorTypeOperations 里
    typedef VectorTypeOperations<T> TypeOperations;

public:
    typedef T ValueType;
    //iterator 直接使用了原生指针
    typedef T* iterator;
    typedef const T* const_iterator;
    typedef std::reverse_iterator<iterator> reverse_iterator;
    typedef std::reverse_iterator<const_iterator>
              const_reverse_iterator;

    Vector()
    {
    }

    // Unlike in std::vector, this constructor does not initialize POD types.
    explicit Vector(size_t size)
        : Base(size, size)
    {
        asanSetInitialBufferSizeTo(size);

        if (begin())
            TypeOperations::initialize(begin(), end());
    }

    Vector(size_t size, const T& val)
        : Base(size, size)
    {
```

```cpp
        asanSetInitialBufferSizeTo(size);

        if (begin())
            TypeOperations::uninitializedFill(begin(), end(), val);
    }

    Vector(std::initializer_list<T> initializerList)
    {
        reserveInitialCapacity(initializerList.size());

        asanSetInitialBufferSizeTo(initializerList.size());

        for (const auto& element : initializerList)
            uncheckedAppend(element);
    }

    ~Vector()
    {
        if (m_size)
            TypeOperations::destruct(begin(), end());

        asanSetBufferSizeToFullCapacity(0);
    }

    Vector(const Vector&);
    template<size_t otherCapacity, typename otherOverflowBehaviour, size_t otherMinimumCapacity, typename OtherMalloc>
    explicit Vector(const Vector<T, otherCapacity, otherOverflowBehaviour, otherMinimumCapacity, OtherMalloc>&);

    Vector& operator=(const Vector&);
    template<size_t otherCapacity, typename otherOverflowBehaviour, size_t otherMinimumCapacity, typename OtherMalloc>
    Vector& operator=(const Vector<T, otherCapacity, otherOverflowBehaviour, otherMinimumCapacity, OtherMalloc>&);

    Vector(Vector&&);
    Vector& operator=(Vector&&);

    //返回 Vector 中元素的个数
    size_t size() const { return m_size; }
    static ptrdiff_t sizeMemoryOffset() { return OBJECT_OFFSETOF(Vector, m_size); }

    //返回的是 Vector 中的容量,容量随着元素的增加、减少而变化
    size_t capacity() const { return Base::capacity(); }
    bool isEmpty() const { return !size(); }

    //这里提供的是数组的访问功能
    T& at(size_t i)
    {
        if (UNLIKELY(i >= size()))
            OverflowHandler::overflowed();
        return Base::buffer()[i];
```

```cpp
}
const T& at(size_t i) const
{
    if (UNLIKELY(i >= size()))
        OverflowHandler::overflowed();
    return Base::buffer()[i];
}
T& at(Checked<size_t> i)
{
    RELEASE_ASSERT(i < size());
    return Base::buffer()[i];
}
const T& at(Checked<size_t> i) const
{
    RELEASE_ASSERT(i < size());
    return Base::buffer()[i];
}
//返回数组中第几个元素
T& operator[](size_t i) { return at(i); }
const T& operator[](size_t i) const { return at(i); }
T& operator[](Checked<size_t> i) { return at(i); }
const T& operator[](Checked<size_t> i) const { return at(i); }

T* data() { return Base::buffer(); }
const T* data() const { return Base::buffer(); }
static ptrdiff_t dataMemoryOffset() { return Base::bufferMemoryOffset(); }
//迭代功能的实现,获取 begin 和 end, Vector 元素有了插入和删除操作
//需要重新 begin 和 end
iterator begin() { return data(); }
iterator end() { return begin() + m_size; }
const_iterator begin() const { return data(); }
const_iterator end() const { return begin() + m_size; }

reverse_iterator rbegin() { return reverse_iterator(end()); }
reverse_iterator rend() { return reverse_iterator(begin()); }
const_reverse_iterator rbegin() const {
    return const_reverse_iterator(end());
}
const_reverse_iterator rend() const {
    return const_reverse_iterator(begin());
}

T& first() { return at(0); }
const T& first() const { return at(0); }
T& last() { return at(size() - 1); }
const T& last() const { return at(size() - 1); }

T takeLast()
{
    T result = WTFMove(last());
    removeLast();
    return result;
}
```

```cpp
        //O(n) 遍历查找的操作，数据量大时使用 HashTable 进行查找的效果会更好
        template<typename U> bool contains(const U&) const;
        template<typename U> size_t find(const U&) const;
        template<typename MatchFunction> size_t findMatching(const MatchFunction&) const;
        template<typename U> size_t reverseFind(const U&) const;

        template<typename U> bool appendIfNotContains(const U&);

        //实现 STL 里的方法，这里的 insert、append 和 resize 等操作会通过
        //reserveCapacity 或 tryReserveCapacity 进行空间扩展实现动态数组
        void shrink(size_t size);
        void grow(size_t size);
        void resize(size_t size);
        void resizeToFit(size_t size);
        void reserveCapacity(size_t newCapacity);
        bool tryReserveCapacity(size_t newCapacity);
        void reserveInitialCapacity(size_t initialCapacity);
        void shrinkCapacity(size_t newCapacity);
        void shrinkToFit() { shrinkCapacity(size()); }

        void clear() { shrinkCapacity(0); }

        void append(ValueType&& value) {
            append<ValueType>(std::forward<ValueType>(value));
        }
        template<typename U> void append(U&&);
        template<typename... Args> void constructAndAppend(Args&&...);
        template<typename... Args> bool tryConstructAndAppend(Args&&...);

        void uncheckedAppend(ValueType&& value) {
            uncheckedAppend<ValueType>(std::forward<ValueType>(value));
        }
        template<typename U> void uncheckedAppend(U&&);

        template<typename U> void append(const U*, size_t);
        template<typename U, size_t otherCapacity> void appendVector(const Vector<U, otherCapacity>&);
        template<typename U> bool tryAppend(const U*, size_t);

        template<typename U> void insert(size_t position, const U*, size_t);
        template<typename U> void insert(size_t position, U&&);
        template<typename U, size_t c> void insertVector(size_t position, const Vector<U, c>&);

        void remove(size_t position);
        void remove(size_t position, size_t length);
        template<typename U> bool removeFirst(const U&);
        template<typename MatchFunction> bool removeFirstMatching(const MatchFunction&, size_t startIndex = 0);
        template<typename U> unsigned removeAll(const U&);
        template<typename MatchFunction> unsigned removeAllMatching(const MatchFunction&, size_t startIndex = 0);
```

```cpp
    void removeLast()
    {
        if (UNLIKELY(isEmpty()))
            OverflowHandler::overflowed();
        shrink(size() - 1);
    }

    void fill(const T&, size_t);
    void fill(const T& val) { fill(val, size()); }

    template<typename Iterator> void appendRange(Iterator start
        , Iterator end);

    MallocPtr<T> releaseBuffer();

    void swap(Vector<T, inlineCapacity, OverflowHandler
        , minCapacity>& other)
    {
#if ASAN_ENABLED
        if (this == std::addressof(other)) // ASan will crash if we try to restrict access to the same buffer twice
            return;
#endif

        // Make it possible to copy inline buffers
        asanSetBufferSizeToFullCapacity();
        other.asanSetBufferSizeToFullCapacity();

        Base::swap(other, m_size, other.m_size);
        std::swap(m_size, other.m_size);

        asanSetInitialBufferSizeTo(m_size);
        other.asanSetInitialBufferSizeTo(other.m_size);
    }

    void reverse();

    void checkConsistency();

    template<typename MapFunction, typename R = typename std::result_of<MapFunction(const T&)>::type> Vector<R> map(MapFunction) const;

private:
    void expandCapacity(size_t newMinCapacity);
    T* expandCapacity(size_t newMinCapacity, T*);
    bool tryExpandCapacity(size_t newMinCapacity);
    const T* tryExpandCapacity(size_t newMinCapacity, const T*);
    template<typename U> U* expandCapacity(size_t newMinCapacity, U*);
    template<typename U> void appendSlowCase(U&&);
    template<typename... Args> void constructAndAppendSlowCase(
        Args&&...);
    template<typename... Args> bool tryConstructAndAppendSlowCase(
        Args&&...);
```

```cpp
    void asanSetInitialBufferSizeTo(size_t);
    void asanSetBufferSizeToFullCapacity(size_t);
    void asanSetBufferSizeToFullCapacity() {
        asanSetBufferSizeToFullCapacity(size());
    }

    void asanBufferSizeWillChangeTo(size_t);

    using Base::m_size;
    using Base::buffer;
    using Base::capacity;
    using Base::swap;
    using Base::allocateBuffer;
    using Base::deallocateBuffer;
    using Base::tryAllocateBuffer;
    using Base::shouldReallocateBuffer;
    using Base::reallocateBuffer;
    using Base::restoreInlineBufferIfNeeded;
    using Base::releaseBuffer;
#if ASAN_ENABLED
    using Base::endOfBuffer;
#endif
};
```

实现 HashTable 的代码如下：

```cpp
    template<typename Key, typename Value, typename Extractor, typename HashFunctions, typename Traits, typename KeyTraits>
    class HashTable {
    public:
        typedef HashTableIterator<Key, Value, Extractor, HashFunctions, Traits, KeyTraits> iterator;
        typedef HashTableConstIterator<Key, Value, Extractor, HashFunctions, Traits, KeyTraits> const_iterator;
        typedef Traits ValueTraits;
        typedef Key KeyType;
        typedef Value ValueType;
        typedef IdentityHashTranslator<ValueTraits, HashFunctions> IdentityTranslatorType;
        typedef HashTableAddResult<iterator> AddResult;

#if DUMP_HASHTABLE_STATS_PER_TABLE
        struct Stats {
            Stats()
                : numAccesses(0)
                , numRehashes(0)
                , numRemoves(0)
                , numReinserts(0)
                , maxCollisions(0)
                , numCollisions(0)
                , collisionGraph()
            {
            }
```

```cpp
            unsigned numAccesses;
            unsigned numRehashes;
            unsigned numRemoves;
            unsigned numReinserts;

            unsigned maxCollisions;
            unsigned numCollisions;
            unsigned collisionGraph[4096];

            void recordCollisionAtCount(unsigned count)
            {
                if (count > maxCollisions)
                    maxCollisions = count;
                numCollisions++;
                collisionGraph[count]++;
            }

            void dumpStats()
            {
                dataLogF("\nWTF::HashTable::Stats dump\n\n");
                dataLogF("%d accesses\n", numAccesses);
                dataLogF("%d total collisions, average %.2f probes per access\n",
numCollisions, 1.0 * (numAccesses + numCollisions) / numAccesses);
                dataLogF("longest collision chain: %d\n", maxCollisions);
                for (unsigned i = 1; i <= maxCollisions; i++) {
                    dataLogF("  %d lookups with exactly %d collisions (%.2f%% , %.2f%%
with this many or more)\n", collisionGraph[i], i, 100.0 * (collisionGraph[i] -
collisionGraph[i+1]) / numAccesses, 100.0 * collisionGraph[i] / numAccesses);
                }
                dataLogF("%d rehashes\n", numRehashes);
                dataLogF("%d reinserts\n", numReinserts);
            }
        };
    #endif

        HashTable();
        ~HashTable()
        {
            invalidateIterators();
            if (m_table)
                deallocateTable(m_table, m_tableSize);
#if CHECK_HASHTABLE_USE_AFTER_DESTRUCTION
            m_table = (ValueType*)(uintptr_t)0xbbadbeef;
#endif
        }

        HashTable(const HashTable&);
        void swap(HashTable&);
        HashTable& operator=(const HashTable&);

        HashTable(HashTable&&);
        HashTable& operator=(HashTable&&);

        // When the hash table is empty,
```

```cpp
            //just return the same iterator for end as for begin.
            // This is more efficient because we don't have to skip all the
            // empty and deleted buckets, and iterating an empty table is a common
            // case that's worth optimizing.
            iterator begin() { return isEmpty() ? end() : makeIterator(m_table); }
            iterator end() { return makeKnownGoodIterator(m_table + m_tableSize); }
            const_iterator begin() const { return isEmpty() ? end() :
makeConstIterator(m_table); }
            const_iterator end() const { return makeKnownGoodConstIterator(m_table +
m_tableSize); }

            unsigned size() const { return m_keyCount; }
            unsigned capacity() const { return m_tableSize; }
            bool isEmpty() const { return !m_keyCount; }

            AddResult add(const ValueType& value) { return
add<IdentityTranslatorType>(Extractor::extract(value), value); }
            AddResult add(ValueType&& value) { return
add<IdentityTranslatorType>(Extractor::extract(value), WTFMove(value)); }

            // A special version of add() that finds the object by
            // hashing and comparing
            // with some other type, to avoid the cost of type conversion
            // if the object is already
            // in the table.
            template<typename HashTranslator, typename T, typename Extra> AddResult
add(T&& key, Extra&&);
            template<typename HashTranslator, typename T, typename Extra> AddResult
addPassingHashCode(T&& key, Extra&&);

            iterator find(const KeyType& key) { return
find<IdentityTranslatorType>(key); }
            const_iterator find(const KeyType& key) const { return
find<IdentityTranslatorType>(key); }
            bool contains(const KeyType& key) const { return
contains<IdentityTranslatorType>(key); }

            template<typename HashTranslator, typename T> iterator find(const T&);
            template<typename HashTranslator, typename T> const_iterator find(const T&)
const;
            template<typename HashTranslator, typename T> bool contains(const T&)
const;

            void remove(const KeyType&);
            void remove(iterator);
            void removeWithoutEntryConsistencyCheck(iterator);
            void removeWithoutEntryConsistencyCheck(const_iterator);
            template<typename Functor>
            void removeIf(const Functor&);
            void clear();

            static bool isEmptyBucket(const ValueType& value) { return
isHashTraitsEmptyValue<KeyTraits>(Extractor::extract(value)); }
            static bool isDeletedBucket(const ValueType& value) { return
```

```cpp
KeyTraits::isDeletedValue(Extractor::extract(value)); }
        static bool isEmptyOrDeletedBucket(const ValueType& value) { return
isEmptyBucket(value) || isDeletedBucket(value); }

        ValueType* lookup(const Key& key) { return
lookup<IdentityTranslatorType>(key); }
        template<typename HashTranslator, typename T> ValueType* lookup(const T&);
        template<typename HashTranslator, typename T> ValueType*
inlineLookup(const T&);

    #if !ASSERT_DISABLED
        void checkTableConsistency() const;
    #else
        static void checkTableConsistency() { }
    #endif
    #if CHECK_HASHTABLE_CONSISTENCY
        void internalCheckTableConsistency() const { checkTableConsistency(); }
        void internalCheckTableConsistencyExceptSize() const
{ checkTableConsistencyExceptSize(); }
    #else
        static void internalCheckTableConsistencyExceptSize() { }
        static void internalCheckTableConsistency() { }
    #endif

    private:
        static ValueType* allocateTable(unsigned size);
        static void deallocateTable(ValueType* table, unsigned size);

        typedef std::pair<ValueType*, bool> LookupType;
        typedef std::pair<LookupType, unsigned> FullLookupType;

        LookupType lookupForWriting(const Key& key) { return
lookupForWriting<IdentityTranslatorType>(key); };
        template<typename HashTranslator, typename T> FullLookupType
fullLookupForWriting(const T&);
        template<typename HashTranslator, typename T> LookupType
lookupForWriting(const T&);

        template<typename HashTranslator, typename T, typename Extra> void
addUniqueForInitialization(T&& key, Extra&&);

        template<typename HashTranslator, typename T> void checkKey(const T&);

        void removeAndInvalidateWithoutEntryConsistencyCheck(ValueType*);
        void removeAndInvalidate(ValueType*);
        void remove(ValueType*);

        bool shouldExpand() const { return (m_keyCount + m_deletedCount) *
m_maxLoad >= m_tableSize; }
        bool mustRehashInPlace() const { return m_keyCount * m_minLoad < m_tableSize
* 2; }
        bool shouldShrink() const { return m_keyCount * m_minLoad < m_tableSize &&
m_tableSize > KeyTraits::minimumTableSize; }
        ValueType* expand(ValueType* entry = nullptr);
```

```cpp
        void shrink() { rehash(m_tableSize / 2, nullptr); }

        ValueType* rehash(unsigned newTableSize, ValueType* entry);
        ValueType* reinsert(ValueType&&);

        static void initializeBucket(ValueType& bucket);
        static void deleteBucket(ValueType& bucket)
{ hashTraitsDeleteBucket<Traits>(bucket); }

        FullLookupType makeLookupResult(ValueType* position, bool found,
          unsigned hash)
            { return FullLookupType(LookupType(position, found), hash); }

        iterator makeIterator(ValueType* pos) { return iterator(this, pos, m_table
+ m_tableSize); }
        const_iterator makeConstIterator(ValueType* pos) const { return
const_iterator(this, pos, m_table + m_tableSize); }
        iterator makeKnownGoodIterator(ValueType* pos) { return iterator(this, pos,
m_table + m_tableSize, HashItemKnownGood); }
        const_iterator makeKnownGoodConstIterator(ValueType* pos) const { return
const_iterator(this, pos, m_table + m_tableSize, HashItemKnownGood); }

    #if !ASSERT_DISABLED
        void checkTableConsistencyExceptSize() const;
    #else
        static void checkTableConsistencyExceptSize() { }
    #endif

    #if CHECK_HASHTABLE_ITERATORS
        void invalidateIterators();
    #else
        static void invalidateIterators() { }
    #endif

        static const unsigned m_maxLoad = 2;
        static const unsigned m_minLoad = 6;

        ValueType* m_table;
        unsigned m_tableSize;
        unsigned m_tableSizeMask;
        unsigned m_keyCount;
        unsigned m_deletedCount;

    #if CHECK_HASHTABLE_ITERATORS
    public:
        // All access to m_iterators should be guarded with m_mutex.
        mutable const_iterator* m_iterators;
        // Use std::unique_ptr so HashTable can still be memmove'd or memcpy'ed.
        mutable std::unique_ptr<Lock> m_mutex;
    #endif

    #if DUMP_HASHTABLE_STATS_PER_TABLE
    public:
        mutable std::unique_ptr<Stats> m_stats;
```

```
#endif
};
```

完整的具体实现可以查看 WTF/Source/wtf/HashTable.h。这里需要注意的是 HashTable 是线性存储空间的起始地址，是用 begin 和 end 表示的。当 capacity 存储空间不足时，WTF 会使用 doubleHash 方法进行扩容，将原存储空间容量扩展两倍。调用 begin 或 end 时，WTF 会创建 HashTable 的 const_iterator，它维持了一个指向 value_type 的指针，所以 iterator 的自增和自减就是指针的自增和自减，这比 STL 的 HashMap 采用数组加链表的实现要简单。

线程

WebKit 对线程的处理就是在一个 loop 循环里处理消息队列，在 WTF/Source/wtf/MessageQueue.h 里有对消息队列的定义，如下所示。

```
// The queue takes ownership of messages and transfer it to the new owner
// when messages are fetched from the queue.
// Essentially, MessageQueue acts as a queue of std::unique_ptr<DataType>.
template<typename DataType>
class MessageQueue {
    WTF_MAKE_NONCOPYABLE(MessageQueue);
public:
    MessageQueue() : m_killed(false) { }
    ~MessageQueue();

    void append(std::unique_ptr<DataType>);
    void appendAndKill(std::unique_ptr<DataType>);
    bool appendAndCheckEmpty(std::unique_ptr<DataType>);
    void prepend(std::unique_ptr<DataType>);

    std::unique_ptr<DataType> waitForMessage();
    std::unique_ptr<DataType> tryGetMessage();
    Deque<std::unique_ptr<DataType>> takeAllMessages();
    std::unique_ptr<DataType> tryGetMessageIgnoringKilled();
    template<typename Predicate>
    std::unique_ptr<DataType> waitForMessageFilteredWithTimeout(
      MessageQueueWaitResult&, Predicate&&, WallTime absoluteTime);

    template<typename Predicate>
    void removeIf(Predicate&&);

    void kill();
    bool killed() const;

    // The result of isEmpty() is only valid if no other thread is
    // manipulating the queue at the same time.
    bool isEmpty();

private:
    //m_mutex 是访问 Deque 的互斥锁，是对 pthread_mutex_t 类型的一个封装
    mutable Lock m_mutex;

    //Condition 是对 pthread_cond_t 类型的封装，m_condition 提供了挂起线程
```

```
    Condition m_condition;

    //内部主要存储结构
    Deque<std::unique_ptr<DataType>> m_queue;
    bool m_killed;
};
```

可以看出 MessageQueue 是通过 pthread_mutex_t 来保证线程安全的。WebKit 在运行时会有很多线程，有网络资源加载的线程、解析页面布局的线程、绘制线程、I/O 线程、解码线程等。最核心的是解析页面布局的线程，其他线程都是由它触发的，所以称其为"主线程"。这个线程在 Mac 端是由 MainThread 定义和实现的；在 iOS 端是由 WebCoreThread 定义和实现的。有些定义比如 callOnMainThread 就是其他异步线程做回调时调用的函数。

3.4.4 Loader

Loader 的主要作用是加载资源，分为 MainResourceLoader 和 SubResourceLoader。加载的资源来自网络、本地或者缓存。Loader 模块本身与平台无关，只是将需要获得资源的请求传给平台相关的网络模块，并且接受网络模块返回的资源。平台相关的网络模块在 WebCore/platform/network 里。如果是 iOS，就在 WebCore/platform/network/ iOS 里；如果是 Mac，就在 WebCore/platform/network/mac 里。

Loader 的资源

网页本身就是一种资源，同时网页还需要其他的资源，比如图片、视频、JavaScript 代码等。下面是主要的资源类型。

- HTML：页面主文件。
- JavaScript：单独的文件，或者将 JavaScript 代码直接写在 HTML 代码里。
- CSS：同 JavaScript 一样可以是单独文件，也可以直接写在 HTML 代码里。
- 图片：各种编码图片，比如 JPEG 文件 和 PNG 文件。
- SVG：矢量图片。
- CSS Shader：为 CSS 带来 3D 图形特性。
- 音频视频：多媒体资源，以及视频字幕。
- 字体：自定义的字体。
- XSL：对 XSLT 语言编写的文件支持。

WebKit 里会有不同的类来表示这些资源，它们有共同的基类 CachedResource，如下图所示，其中 HTML 的资源类型是 CachedRawResource，基类为 CachedResource。

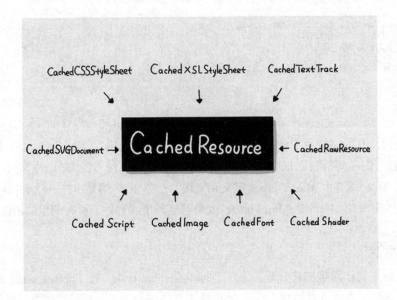

FrameLoader

FrameLoader 是加载页面的入口,最终 FrameLoader 会异步调用 Load,如下图所示。FrameLoader 主要提供了下载和 Frame 的接口。任何一个页面都需要至少一个 MainFrame,所以 FrameLoader 加载页面一般都会先加载一个 MainFrame。

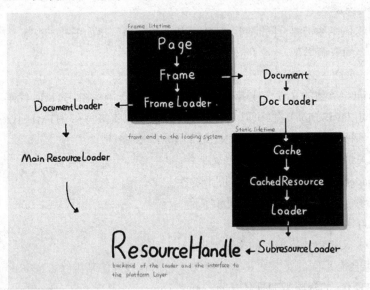

下面是对代码里主要类的说明:

- Frame 和 FrameLoaderClient:在 Frame 的构造函数中调用 FrameLoader 的构造函数需要传入 Frame 指针和 FrameLoaderClient 指针。一般作为不同平台适配代码,控制 WebKit 的使用逻辑。FrameLoader 会把加载过程的状态、结果等信息传递给 FrameLoaderClient,由此 FrameLoaderClient 就能够掌控 FrameLoader 的动作了。
- SubFrameLoader:维护子 Frame。

- DocumentWriter：有一个 m_writer 对象在 Frame 的数据 load 完成时，进行下一步操作，比如解码处理。
- DocumentLoader：FrameLoader 会维护三个 DocumentLoader。m_policyDocumentLoader 作用于 policy check 阶段。m_provisionalDocumentLoader 会负责 startLoadingMainResource 的调用。数据到了后会使用 m_documentLoader，这时前一个 Frame 的 DocumentLoader 则不能再被使用。
- HistoryController：管理历史记录，HistoryItem 用来保存和恢复 Document 与 Page 的状态，维护浏览页面的前进、后退队列，这样就可以实现前进、后退等操作。FrameLoader 通过 HistoryController 来操作 m_backForwardController 对象。
- ResourceLoadNotifier：当 ResourceLoader 有变化时通知 FrameLoader，用于 ResourceLoader 和 FrameLoader 之间的通信。
- SubframeLoader：用来控制 MainFrame 中 iframe 的加载。
- FrameLoaderStateMachine：描述 FrameLoader 处理 DocumentLoader 的节点的状态，看其是在创建状态还是显示状态。
- policyChecker：对 FrameLoader 做校验，有三种类型：NewWindow Policy 指对新开的 tab 或 window 的校验。NavigationPolicy 指对页面请求的校验。ContentPolicy 指对请求收到的数据等的校验。policyChecker 会提供对应接口，由 FrameLoaderClient 对这些请求进行校验以确定是否继续加载或者做其他操作。
- MainResourceLoader：加载资源，包含 HTML，以及 CSS、JavaScript 和 Image 等这些是由 SubResourceLoader 管理的资源。
- CacheResourceLoader：读取缓存资源。资源加载和缓存是由 ResourceLoader 和 Cache 这两个独立性高的模块来实现的。
- DocumentWriter：辅助类，将获取的数据写到 writer 里，会创建 DOM Tree 的根节点 HTMLDocument 对象，作为一个缓冲区进行 DocumentParser 解析操作，同时该类还包含了文档字符解码类和 HTMLDocumentParser 类。

解析前创建不同的 Dcoument。DocumentLoader 的 commitData 在第一次收到数据时会调用 DcumentWriter 的 begin。这个 begin 再调用 DocumentWriter::createDocument 创建一个 Document 对象。

创建的对象会根据 URL 来创建不同的 Document。创建出来的 Document 有 HTMLDocument，详细实现可以参考 DOMImplementation::createDocument 的实现。

FrameLoader 的主要接口

Frame::init 是 FrameLoader 自身的初始化。初始化的调用顺序如下：

```
WebFrame::WebFrame(webPage* parent,WebFrameData *frameData)
WebFramePrivate::init(WebFrame* webframe,WebFrameData* frameData)
Frame::init()
FrameLoader::init()
```

FrameLoader::commitProvisionalLoad 在提交 provisional 阶段下载数据，完成 Document loading 的调用顺序。

```
DocumentLoader::finishLoading
DocumentLoader::commitIfReady
FrameLoader::commitProvisionalLoad
//资源数据接受提交调用顺序
ResourceLoader::didReceiveData
MainResourceLoader::addData
DocumentLoader::receiveData
DocumentLoader::commitLoad
DocumentLoader::commitIfReady
DocumentLoader::commitProvisionalLoad
```

Frame::finishedLoading 是网络加载完成的接口,作用是通知 DocumentLoader 和 DocumentWriter 任务已经完成,可以进行后面提交数据和解析的工作了,函数的调用顺序如下:

```
ResourceLoader::didFinishLoading
MainResourceLoader::didFinishLoading
FrameLoader::finishedLoading
FrameLoader::init()
```

FrameLoader::finishedParsing 是完成解析时调用的接口,调用顺序如下:

```
DocumentWritter::end
Document::finishParsing
Document::finishedParsing
FrameLoader::finishedParsing
```

FrameLoader::load(FrameLoadRequest&& request) 的作用是加载请求,将 Frame 相关的数据封装成 ResourceRequest,然后将 ResourceRequest 作为参数带入 FrameLoader::Load 接口里,这个接口会创建出 DocumentLoader。

```
//创建 DocumentLoader
void FrameLoader::load(FrameLoadRequest&& request)
{
    if (m_inStopAllLoaders)
        return;

    if (!request.frameName().isEmpty()) {
        Frame* frame = findFrameForNavigation(request.frameName());
        if (frame) {
            request.setShouldCheckNewWindowPolicy(false);
            if (&frame->loader() != this) {
                frame->loader().load(WTFMove(request));
                return;
            }
        }
    }

    if (request.shouldCheckNewWindowPolicy()) {
        NavigationAction action { request.requester(), request.resourceRequest(),
InitiatedByMainFrame::Unknown, NavigationType::Other,
```

```cpp
request.shouldOpenExternalURLsPolicy() };
        policyChecker().checkNewWindowPolicy(WTFMove(action),
request.resourceRequest(), nullptr, request.frameName(), [this] (const
ResourceRequest& request, FormState* formState, const String& frameName, const
NavigationAction& action, bool shouldContinue) {
            continueLoadAfterNewWindowPolicy(request, formState, frameName, action,
shouldContinue, AllowNavigationToInvalidURL::Yes, NewFrameOpenerPolicy::Suppress);
        });

        return;
    }

    if (!request.hasSubstituteData())
        request.setSubstituteData(defaultSubstituteDataForURL(request
          .resourceRequest().url()));
    //FrameLoader 使用空 SubstituteData 来创建 DocumentLoader
    //完成 MainResource 的加载
    Ref<DocumentLoader> loader = m_client.createDocumentLoader(
request.resourceRequest(), request.substituteData());
    applyShouldOpenExternalURLsPolicyToNewDocumentLoader(m_frame, loader,
request);

    load(loader.ptr());
}

//完善 request 信息
void FrameLoader::load(DocumentLoader* newDocumentLoader)
{
    ResourceRequest& r = newDocumentLoader->request();
    //ResourceRequest 包含了 HTTP Header 的内容
    //addExtraFieldsToMainResourceRequest 这个方法会添加一些类似 Cookie 策略
    //User Agent、Cache-Control、Content-type 等信息
    addExtraFieldsToMainResourceRequest(r);
    FrameLoadType type;

    if (shouldTreatURLAsSameAsCurrent(newDocumentLoader
      ->originalRequest().url())) {
        r.setCachePolicy(ReloadIgnoringCacheData);
        type = FrameLoadType::Same;
    } else if (shouldTreatURLAsSameAsCurrent(newDocumentLoader
      ->unreachableURL()) && m_loadType == FrameLoadType::Reload)
        type = FrameLoadType::Reload;
    else if (m_loadType == FrameLoadType
      ::RedirectWithLockedBackForwardList
      && !newDocumentLoader->unreachableURL().isEmpty()
      && newDocumentLoader->substituteData().isValid())
        type = FrameLoadType::RedirectWithLockedBackForwardList;
    else
        type = FrameLoadType::Standard;

    if (m_documentLoader)
        newDocumentLoader->setOverrideEncoding(m_documentLoader
          ->overrideEncoding());
```

```cpp
    if (shouldReloadToHandleUnreachableURL(newDocumentLoader)) {
        history().saveDocumentAndScrollState();
        ASSERT(type == FrameLoadType::Standard);
        type = FrameLoadType::Reload;
    }
    loadWithDocumentLoader(newDocumentLoader, type, 0
        , AllowNavigationToInvalidURL::Yes);
}

//校验检查
void FrameLoader::loadWithDocumentLoader(DocumentLoader* loader, FrameLoadType
type, FormState* formState, AllowNavigationToInvalidURL allowNavigationToInvalidURL)
{
    // Retain because dispatchBeforeLoadEvent
    // may release the last reference to it.
    Ref<Frame> protect(m_frame);

    ASSERT(m_client.hasWebView());
    ASSERT(m_frame.view());

    if (!isNavigationAllowed())
        return;

    if (m_frame.document())
        m_previousURL = m_frame.document()->url();

    const URL& newURL = loader->request().url();
    // Log main frame navigation types.
    if (m_frame.isMainFrame()) {
        if (auto* page = m_frame.page())
            page->mainFrameLoadStarted(newURL, type);
            static_cast<MainFrame&>(m_frame).performanceLogging()
                .didReachPointOfInterest(PerformanceLogging
                ::MainFrameLoadStarted);
    }

    policyChecker().setLoadType(type);
    bool isFormSubmission = formState;

    const String& httpMethod = loader->request().httpMethod();

    if (shouldPerformFragmentNavigation(isFormSubmission, httpMethod,
policyChecker().loadType(), newURL)) {
        RefPtr<DocumentLoader> oldDocumentLoader = m_documentLoader;
        NavigationAction action { *m_frame.document(), loader->request(),
InitiatedByMainFrame::Unknown, policyChecker().loadType(), isFormSubmission };

        oldDocumentLoader->setTriggeringAction(action);
        oldDocumentLoader->setLastCheckedRequest(ResourceRequest());
        policyChecker().stopCheck();
        policyChecker().checkNavigationPolicy(loader->request(), false /*
didReceiveRedirectResponse */, oldDocumentLoader.get(), formState, [this] (const
ResourceRequest& request, FormState*, bool shouldContinue) {
            continueFragmentScrollAfterNavigationPolicy(request,
```

```cpp
shouldContinue);
        });
        return;
    }

    if (Frame* parent = m_frame.tree().parent())
loader->setOverrideEncoding(parent->loader().documentLoader()->overrideEncoding());

    policyChecker().stopCheck();

    //把 DocumentLoader 赋值给 m_policyDocumentLoader
    setPolicyDocumentLoader(loader);

    //将请求信息记在 loader.m_triggeringAction 中
    if (loader->triggeringAction().isEmpty())
        loader->setTriggeringAction({ *m_frame.document(), loader->request(),
InitiatedByMainFrame::Unknown, policyChecker().loadType(), isFormSubmission });

    if (Element* ownerElement = m_frame.ownerElement()) {
        if (!m_stateMachine.committedFirstRealDocumentLoad()
&& !ownerElement->dispatchBeforeLoadEvent(loader->request().url().string())) {
            continueLoadAfterNavigationPolicy(loader->request(), formState, false,
allowNavigationToInvalidURL);
            return;
        }
    }
    //使用 loader.m_triggeringAction 做校验、处理空白等,用来决定如何处理请求
        policyChecker().checkNavigationPolicy(loader->request(), false /*
didReceiveRedirectResponse */, loader, formState, [this, allowNavigationToInvalidURL]
(const ResourceRequest& request, FormState* formState, bool shouldContinue) {
        //不同的 shouldContinue 的流程也会不同, formState 是用来判断 HTMLFormElement 表单
        //FormSubmissionTrigger 枚举状态的
        continueLoadAfterNavigationPolicy(request, formState, shouldContinue,
allowNavigationToInvalidURL);
    });
}

    //
    void FrameLoader::continueLoadAfterNavigationPolicy(const ResourceRequest&
request, FormState* formState, bool shouldContinue, AllowNavigationToInvalidURL
allowNavigationToInvalidURL)
    {
        // If we loaded an alternate page to replace an unreachableURL,
        // we'll get in here with a nil policyDataSource
        // because loading the alternate page will have passed
        // through this method already, nested;
        // otherwise, policyDataSource should still be set.
        ASSERT(m_policyDocumentLoader || !m_provisionalDocumentLoader
          ->unreachableURL().isEmpty());

        bool isTargetItem = history().provisionalItem() ?
```

```cpp
history().provisionalItem()->isTargetItem() : false;

    bool urlIsDisallowed = allowNavigationToInvalidURL ==
AllowNavigationToInvalidURL::No && !request.url().isValid();

    // Three reasons we can't continue:
    //   1) Navigation policy delegate said we can't so request is nil.
    //      A primary case of this
    //      is the user responding Cancel to the form repost nag sheet.
    //   2) User responded Cancel to an alert popped up by the before
    //      unload event handler.
    //   3) The request's URL is invalid and navigation to invalid URLs
    //      is disallowed.
    bool canContinue = shouldContinue && shouldClose() && !urlIsDisallowed;

    if (!canContinue) {
        // If we were waiting for a quick redirect,
        // but the policy delegate decided to ignore it, then we
        // need to report that the client redirect was cancelled.
        // FIXME: The client should be told about ignored
        // non-quick redirects, too.
        if (m_quickRedirectComing)
            clientRedirectCancelledOrFinished(false);

        setPolicyDocumentLoader(nullptr);

        // If the navigation request came from the back/forward menu,
        // and we punt on it, we have the
        // problem that we have optimistically moved the b/f cursor already,
        // so move it back. For sanity,
        // we only do this when punting a navigation for the
        // target frame or top-level frame.
        if ((isTargetItem || m_frame.isMainFrame()) &&
          isBackForwardLoadType(policyChecker().loadType())) {
            if (Page* page = m_frame.page()) {
                if (HistoryItem* resetItem = m_frame.mainFrame().loader()
                  .history().currentItem()) {
                    page->backForward().setCurrentItem(resetItem);
                    m_frame.loader().client()
                      .updateGlobalHistoryItemForPage();
                }
            }
        }
        return;
    }

    FrameLoadType type = policyChecker().loadType();
    // A new navigation is in progress,
    // so don't clear the history's provisional item.
    stopAllLoaders(ShouldNotClearProvisionalItem);

    // <rdar://problem/6250856> - In certain circumstances on pages with multiple
frames, stopAllLoaders()
    // might detach the current FrameLoader,
```

```
   // in which case we should bail on this newly defunct load.
   if (!m_frame.page())
      return;

   //把 DocumentLoader 赋值给 m_provisionalDocumentLoader
   setProvisionalDocumentLoader(m_policyDocumentLoader.get());
   m_loadType = type;
   //设置 FrameLoader 状态为 Provisional
   setState(FrameStateProvisional);

   setPolicyDocumentLoader(nullptr);

   if (isBackForwardLoadType(type)) {
      auto& diagnosticLoggingClient = m_frame.page()
                                   ->diagnosticLoggingClient();
      if (history().provisionalItem()->isInPageCache()) {
         diagnosticLoggingClient.logDiagnosticMessageWithResult(
            DiagnosticLoggingKeys::pageCacheKey(), DiagnosticLoggingKeys
            ::retrievalKey(), DiagnosticLoggingResultPass,
            ShouldSample::Yes);
         loadProvisionalItemFromCachedPage();
         return;
      }
      diagnosticLoggingClient.logDiagnosticMessageWithResult(
         DiagnosticLoggingKeys::pageCacheKey(), DiagnosticLoggingKeys
         ::retrievalKey(), DiagnosticLoggingResultFail,
         ShouldSample::Yes);
   }

   if (!formState) {
      continueLoadAfterWillSubmitForm();
      return;
   }

   m_client.dispatchWillSubmitForm(*formState, [this] (
     PolicyAction action) {
      policyChecker().continueLoadAfterWillSubmitForm(action);
   });
}
```

接着，在 continueLoadAfterWillSubmitForm()方法里 m_provisionalDocumentLoader 会做些准备工作，并且将状态设置成为正在加载资源。在 prepareForLoadStart 中抛出加载状态，回调给 FrameLoaderClient，调用 DocumentLoader 的 startLoadingMainResource。

WebKit 网络处理

WebKit 资源的加载都是由其移植方实现的，网络部分代码在 WebCore/platform/network 里，包含了 HTTP Header、MIME、状态码等信息的处理。

ResourceHandle 的相关类

- ResouceHandleClient：与网络传输过程相关，是网络事件对应的回调，如下图所示。

- MainResourceLoader：主资源，读取 HTML 页面会由 MainResourceLoader 类管理，下载失败时会向用户报错，没有缓存。
- SubResourceLoader：派生资源，在 HTML 页面里内嵌图片和脚本链接等，通过 ResourceScheduler 类来管理资源，加载调度。下载失败时不向用户报错。有缓存机制，通过 ResourceCache 保存 CSS、JavaScript 等资源原始数据，以及解码后的图片数据，一方面可以节省流量，另一方面可以缩短解码时间。
- ResourceLoader：主资源和派生资源加载器的基类，通过 ResourceNotifier 将资源通过回调传给 FrameLoaderClient。

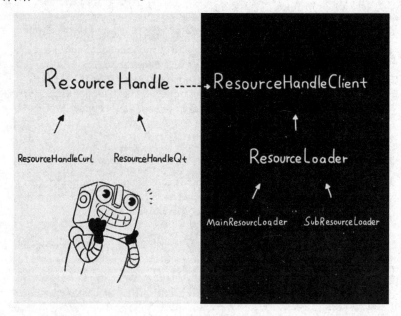

ResourceHandleClient

ResourceHandleClient 定义了对应网络加载事件回调处理的虚函数，如下所示。

```
class ResourceHandleClient {
public:
    WEBCORE_EXPORT ResourceHandleClient();
    WEBCORE_EXPORT virtual ~ResourceHandleClient();

    WEBCORE_EXPORT virtual ResourceRequest willSendRequest(
        ResourceHandle*, ResourceRequest&&, ResourceResponse&&);
    virtual void didSendData(ResourceHandle*, unsigned long long /*bytesSent*/,
unsigned long long /*totalBytesToBeSent*/) { }

    //收到服务器端第一个响应包，通过里面的 HTTP Header 可以判断是否成功
    virtual void didReceiveResponse(ResourceHandle*, ResourceResponse&&) { }
    //收到服务器端包含请求数据的响应包
    virtual void didReceiveData(ResourceHandle*, const char*, unsigned, int /*encodedDataLength*/) { }
    WEBCORE_EXPORT virtual void didReceiveBuffer(ResourceHandle*,
Ref<SharedBuffer>&&, int encodedDataLength);
    //接受过程结束
```

```cpp
        virtual void didFinishLoading(ResourceHandle*) { }
        //接受失败
        virtual void didFail(ResourceHandle*, const ResourceError&) { }
        virtual void wasBlocked(ResourceHandle*) { }
        virtual void cannotShowURL(ResourceHandle*) { }

        virtual bool usesAsyncCallbacks() { return false; }

        virtual bool loadingSynchronousXHR() { return false; }

        // Client will pass an updated request using
        // ResourceHandle::continueWillSendRequest() when ready.
        WEBCORE_EXPORT virtual void willSendRequestAsync(ResourceHandle*,
ResourceRequest&&, ResourceResponse&&);

        // Client will call ResourceHandle::continueDidReceiveResponse()
        // when ready.
        WEBCORE_EXPORT virtual void didReceiveResponseAsync(ResourceHandle*,
ResourceResponse&&);

    #if USE(PROTECTION_SPACE_AUTH_CALLBACK)
        // Client will pass an updated request using
        // ResourceHandle::continueCanAuthenticateAgainstProtectionSpace()
        // when ready.
        WEBCORE_EXPORT virtual void
canAuthenticateAgainstProtectionSpaceAsync(ResourceHandle*, const
ProtectionSpace&);
    #endif
        // Client will pass an updated request using
        // ResourceHandle::continueWillCacheResponse() when ready.
    #if USE(CFURLCONNECTION)
        WEBCORE_EXPORT virtual void willCacheResponseAsync(ResourceHandle*,
CFCachedURLResponseRef);
    #elif PLATFORM(COCOA)
        WEBCORE_EXPORT virtual void willCacheResponseAsync(ResourceHandle*,
NSCachedURLResponse *);
    #endif

    #if USE(SOUP)
        virtual char* getOrCreateReadBuffer(size_t /*requestedLength*/, size_t&
/*actualLength*/) { return 0; }
    #endif

        virtual bool shouldUseCredentialStorage(ResourceHandle*) { return false; }
        virtual void didReceiveAuthenticationChallenge(ResourceHandle*, const
AuthenticationChallenge&) { }
    #if USE(PROTECTION_SPACE_AUTH_CALLBACK)
        virtual bool canAuthenticateAgainstProtectionSpace(ResourceHandle*, const
ProtectionSpace&) { return false; }
    #endif
        virtual void receivedCancellation(ResourceHandle*, const
AuthenticationChallenge&) { }

    #if PLATFORM(IOS) || USE(CFURLCONNECTION)
```

```
        virtual RetainPtr<CFDictionaryRef> connectionProperties(ResourceHandle*)
{ return nullptr; }
    #endif

    #if USE(CFURLCONNECTION)
        virtual CFCachedURLResponseRef willCacheResponse(ResourceHandle*,
CFCachedURLResponseRef response) { return response; }
    #if PLATFORM(WIN)
        virtual bool shouldCacheResponse(ResourceHandle*, CFCachedURLResponseRef)
{ return true; }
    #endif // PLATFORM(WIN)

    #elif PLATFORM(COCOA)
        virtual NSCachedURLResponse *willCacheResponse(ResourceHandle*,
NSCachedURLResponse *response) { return response; }
    #endif
    };
```

加载流程

MainResourceLoader 加载的是 HTML 文本资源，加载的顺序如下：

```
MainResourceLoader::load
MainResourceLoader::loadNow
MainResourceLoader::willSendRequest
ResourceLoader::willSendRequest policyChecker::checkNavigationPolicy
ResourceHandle::create
MainResourceLoader::didReceiveResponse
policyChecker:: checkContentPolicy
MainResourceLoader::continueAfterContentPolicy
ResourceLoader::didReceiveResponse
MainResourceLoader::didReceiveData
ResourceLoader::didReceiveData
MainResourceLoader::addData
DocumentLoader::receivedData
DocumentLoader::commitLoad
FrameLoader::commitProvisionalLoad
FrameLoaderClientQt::committedLoad
DocumentLoader::commitData
DocumentWriter::setEncoding
DocumentWriter::addData
DocumentParser::appendByte
DecodedDataDocumentParser::appendBytes
HTMLDocumentParser::append
MainResourceLoader::didFinishLoading
FrameLoader::finishedLoading
DocumentLoader::finishedLoading
FrameLoader::finishedLoadingDocument
DocumentWriter::end
Document::finishParsing
HTMLDocumentParser::finish
```

FrameLoaderClient 在收到第一个响应包后会回调给 MainResourceLoader 的 didReceiveResponse。函数处理 Header 时使用 policyChecker 的 checkContentPolicy 做校验。当 PolicyAction 为 PolicyUse

时，就回调给 MainResourceLoader 的 continueAfterContentPolicy。ResourceResponse 会通过 ResourceNotifier 将结果通知给外部不同平台的 FrameLoaderClient。FrameLoaderClient 通过回调来接收结果。

FrameLoaderClient 在收到后面的带数据的响应包后，会回调给 MainResourceLoader 的 didReceiveData。通过 ResourceLoader 的 addDataOrBuffer 把收到的数据存到 m_resourceData 里。下面是 addDataOrBuffer 方法的实现代码：

```
void ResourceLoader::addDataOrBuffer(const char* data, unsigned length,
SharedBuffer* buffer, DataPayloadType dataPayloadType)
{
    if (m_options.dataBufferingPolicy == DoNotBufferData)
        return;

    if (!m_resourceData || dataPayloadType == DataPayloadWholeResource) {
        if (buffer)
            m_resourceData = buffer;
        else
            m_resourceData = SharedBuffer::create(data, length);
        return;
    }

    if (buffer)
        m_resourceData->append(*buffer);
    else
        m_resourceData->append(data, length);
}
```

接收到的 receiveData 会进入到 DocumentLoader 的 commitLoad 里。

```
void DocumentLoader::commitLoad(const char* data, int length)
{
    // Both unloading the old page and parsing the new page
    // may execute JavaScript which destroys the datasource
    // by starting a new load, so retain temporarily.
    RefPtr<Frame> protectedFrame(m_frame);
    Ref<DocumentLoader> protectedThis(*this);

    commitIfReady();
    FrameLoader* frameLoader = DocumentLoader::frameLoader();
    if (!frameLoader)
        return;
#if ENABLE(WEB_ARCHIVE) || ENABLE(MHTML)
    if (ArchiveFactory::isArchiveMimeType(response().mimeType()))
        return;
#endif
    //FrameLoader 通过 transitionToCommitted
    //从 Provisional 状态变为 Committed 状态
    frameLoader->client().committedLoad(this, data, length);

    if (isMultipartReplacingLoad())
        frameLoader->client().didReplaceMultipartContent();
}
```

随后 DocumentLoader 执行 commitData，将前面接收到的 receiveData 进行编码。这些数据会通过 DocumentParser 来解码，解码后的数据会创建 HTMLDocument 和 Document 对象，并且通过 DocumentWriter 传给 DocumentParser。整个过程在 DocumentWriter 的 begin 方法里完成，具体实现代码如下：

```cpp
void DocumentWriter::begin(const URL& urlReference, bool dispatch, Document* ownerDocument)
{
    // We grab a local copy of the URL because it's easy for callers to supply
    // a URL that will be deallocated during the execution of this function.
    // For example, see <https://bugs.webkit.org/show_bug.cgi?id=66360>.
    URL url = urlReference;

    // Create a new document before clearing the frame, because it may need to
    // inherit an aliased security context.
    //创建了 Document 对象
    Ref<Document> document = createDocument(url);

    // If the new document is for a Plugin but we're supposed to be
    // sandboxed from Plugins,
    // then replace the document with one whose parser will
    // ignore the incoming data (bug 39323)
    if (document->isPluginDocument() && document->isSandboxed(SandboxPlugins))
        document = SinkDocument::create(m_frame, url);

    // FIXME: Do we need to consult the content security policy here
    // about blocked plug-ins?

    bool shouldReuseDefaultView = m_frame->loader().stateMachine()
        .isDisplayingInitialEmptyDocument() && m_frame->document()
        ->isSecureTransitionTo(url);
    if (shouldReuseDefaultView)
        document->takeDOMWindowFrom(m_frame->document());
    else
        document->createDOMWindow();

    // Per <http://www.w3.org/TR/upgrade-insecure-requests/>,
    // we need to retain an ongoing set of upgraded
    // requests in new navigation contexts.
    // Although this information is present when we construct the
    // Document object, it is discard in the subsequent 'clear'
    // statements below. So, we must capture it
    // so we can restore it.
    HashSet<RefPtr<SecurityOrigin>> insecureNavigationRequestsToUpgrade;
    if (auto* existingDocument = m_frame->document())
        insecureNavigationRequestsToUpgrade = existingDocument
            ->contentSecurityPolicy()->takeNavigationRequestsToUpgrade();

    m_frame->loader().clear(document.ptr(), !shouldReuseDefaultView,
        !shouldReuseDefaultView);
    clear();

    // m_frame->loader().clear() might fire unload event which
```

```
    // could remove the view of the document.
    // Bail out if document has no view.
    if (!document->view())
        return;

    if (!shouldReuseDefaultView)
        m_frame->script().updatePlatformScriptObjects();

    m_frame->loader().setOutgoingReferrer(url);
    m_frame->setDocument(document.copyRef());

    document->contentSecurityPolicy()
      ->setInsecureNavigationRequestsToUpgrade(
      WTFMove(insecureNavigationRequestsToUpgrade));

    if (m_decoder)
        document->setDecoder(m_decoder.get());
    if (ownerDocument) {
        document->setCookieURL(ownerDocument->cookieURL());
        document->setSecurityOriginPolicy(ownerDocument
            ->securityOriginPolicy());
        document->setStrictMixedContentMode(ownerDocument
            ->isStrictMixedContentMode());
    }

    m_frame->loader().didBeginDocument(dispatch);

    document->implicitOpen();

    // We grab a reference to the parser so that we'll always
    // send data to the
    // original parser, even if the document acquires
    // a new parser (e.g., via document.open).
    m_parser = document->parser();

    if (m_frame->view() && m_frame->loader().client().hasHTMLView())
        m_frame->view()->setContentsSize(IntSize());

    m_state = StartedWritingState;
}
```

Document 对象是 DOM Tree 的根节点,m_writer 在 addData 时会用 DocumentParser 来解码。解码过程是在 DecodedDataDocumentParser 的 appendBytes 里完成的。代码如下:

```
void DecodedDataDocumentParser::appendBytes(DocumentWriter& writer, const char* data, size_t length)
{
    if (!length)
        return;

    //解码 Loader 模块传来的字节流
    //解码后的 String 会被 HTMLDocumentParser 的 HTMLInputStream 所持有
    String decoded = writer.createDecoderIfNeeded()->decode(data, length);
    if (decoded.isEmpty())
```

```
        return;
    //解码成功后交给 Parser 处理生成 DOM Tree
    writer.reportDataReceived();
    append(decoded.releaseImpl());
}
```

数据接收完毕后，开始回调 MainResourceLoader 的 didFinishLoading。接下来，再调用 FrameLoader 的 finishedLoading 和 finishedLoadingDocument，以及 DocumentWriter 的 end。最后调用 Document 的 finishParsingMainResource，加载完成。

下面讲解一下 SubResourceLoader 的加载顺序。SubResourceLoader 加载的都是派生资源。在创建 HTMLElement、设置 src 属性时，HTMLConstructionSite 会判断 tagName 是否是 img。如果是 img，就会触发 SubResourceLoader。设置属性和判断的代码在 HTMLConstructionSite 类里。具体实现代码如下：

```
RefPtr<Element> HTMLConstructionSite
  ::createHTMLElementOrFindCustomElementInterface(AtomicHTMLToken
  & token, JSCustomElementInterface** customElementInterface)
{
    auto& localName = token.name();
    // FIXME: This can't use HTMLConstructionSite::createElement
    // because we have to pass the current form element.
    //  We should rework form association to occur after construction to
    // allow better code sharing here.
    // http://www.whatwg.org/specs/web-apps/current-work/multipage
    // /tree-construction.html#create-an-element-for-the-token
    Document& ownerDocument = ownerDocumentForCurrentNode();
    bool insideTemplateElement = !ownerDocument.frame();
    //将 tagName 和节点构造创建成 HTMLImageElement
    RefPtr<Element> element = HTMLElementFactory::createKnownElement(
       localName, ownerDocument, insideTemplateElement ? nullptr
       : form(), true);
    if (UNLIKELY(!element)) {
        auto* window = ownerDocument.domWindow();
        if (customElementInterface && window) {
            auto* registry = window->customElementRegistry();
            if (UNLIKELY(registry)) {
               if (auto* elementInterface = registry
                 ->findInterface(localName)) {
                   *customElementInterface = elementInterface;
                   return nullptr;
               }
            }
        }

        QualifiedName qualifiedName(nullAtom(), localName,
          xhtmlNamespaceURI);
        if (Document::validateCustomElementName(localName) ==
CustomElementNameValidationStatus::Valid) {
            element = HTMLElement::create(qualifiedName, ownerDocument);
            element->setIsCustomElementUpgradeCandidate();
        } else
            element = HTMLUnknownElement::create(qualifiedName,
```

```
            ownerDocument);
    }
    ASSERT(element);

    // FIXME: This is a hack to connect images to pictures before the image has
    // been inserted into the document. It can be removed once asynchronous image
    // loading is working.
    if (is<HTMLPictureElement>(currentNode())
      && is<HTMLImageElement>(*element))
        downcast<HTMLImageElement>(*element).setPictureElement(
          &downcast<HTMLPictureElement>(currentNode()));
    setAttributes(*element, token, m_parserContentPolicy);
    ASSERT(element->isHTMLElement());
    return element;
}
```

在 HTMLImageElement 类里，HTMLImageElement 会用 parseAttribute 方法处理属性。当 parseAttribute 方法发现 HTMLImageElement 的属性是 srcAttr 时，就会触发 selectImageSource 方法。方法实现如下：

```
void HTMLImageElement::selectImageSource()
{
    // First look for the best fit source from our <picture> parent
    // if we have one.
    ImageCandidate candidate = bestFitSourceFromPictureElement();
    if (candidate.isEmpty()) {
        // If we don't have a <picture> or didn't find a source,
        // then we use our own attributes.
        auto sourceSize = SizesAttributeParser(attributeWithoutSynchronization(sizesAttr).string(), document()).length();
        candidate = bestFitSourceForImageAttributes(document().deviceScaleFactor(), attributeWithoutSynchronization(srcAttr), attributeWithoutSynchronization(srcsetAttr), sourceSize);
    }
    setBestFitURLAndDPRFromImageCandidate(candidate);
    m_imageLoader.updateFromElementIgnoringPreviousError();
}
```

SubResourceLoader 的主要处理工作都是交由 SubResourceLoaderClient 类来完成的。ResourceLoadScheduler 会对 SubResourceLoader 进行调度管理。

- Document：创建一个拥有 CachedResourceLoader 类的对象实例 m_cachedResourceLoader。这个类有访问派生资源的接口，比如 requestImage、requestCSSStyleSheet、requestUserCSSStyleSheet、requestScript、requestFont、requestXSLStyleSheet、requestLinkPrefetch 等。requestImage 有条件判断，因此外部实现接口就可以通过设置不加载图片，达到在浏览器里实现无图模式的目的了。
- CachedResourceLoader：创建 CachedResourceRequest 对象发送请求。在它的构造函数里会传入 CachedResource 对象作为参数。通过 m_requests 记录资源请求数量，这样

就可以把加载状态回调出去了。
- MemoryCache：维护一个 HashMap，其中的 value 存储 CachedResource 类缓存的内容。HashMap 的定义为 HashMap <String,CachedResource> m_resources。

在 CachedResourceLoader 里通过 resourceForUrl 在 MemoryCache 里找 URL 对应的缓存资源。determineRevalidationPolicy 会返回一个枚举值用来确定用哪种方式加载资源，如下所示。

```
enum RevalidationPolicy { Use, Revalidate, Reload, Load };
    RevalidationPolicy determineRevalidationPolicy(CachedResource::Type,
CachedResourceRequest&, CachedResource* existingResource, ForPreload, DeferOption)
const;
```

可以看到 RevalidationPolicy 有四种，Use、Revalidate、Reload 和 Load。Use 表示资源已经在缓存里，可以直接使用。Revalidate 表示缓存资源已失效，需删除旧资源，重新在缓存里加新资源。Reload 表示需要重新加载，满足重新加载触发条件首先要资源已缓存，并且资源类型不同，或 Cache-control:no-cache 设置为不使用 Cache，达到条件就会进行重新加载。Load 表示缓存里没有缓存的资源，需要直接加载。这段逻辑处理在 CachedResourceLoader 里的 requestResource 里有实现，具体代码如下：

```
    RevalidationPolicy policy = determineRevalidationPolicy(type, request,
resource.get(), forPreload, defer);
    switch (policy) {
    case Reload:
        memoryCache.remove(*resource);
        FALLTHROUGH;
    case Load:
        if (resource)
            logMemoryCacheResourceRequest(frame(),
DiagnosticLoggingKeys::memoryCacheEntryDecisionKey(),
DiagnosticLoggingKeys::unusedKey());
        resource = loadResource(type, WTFMove(request));
        break;
    case Revalidate:
        if (resource)
            logMemoryCacheResourceRequest(frame(),
DiagnosticLoggingKeys::memoryCacheEntryDecisionKey(),
DiagnosticLoggingKeys::revalidatingKey());
        resource = revalidateResource(WTFMove(request), *resource);
        break;
    case Use:
        ASSERT(resource);
        if (shouldUpdateCachedResourceWithCurrentRequest(*resource, request)) {
            resource = updateCachedResourceWithCurrentRequest(*resource,
WTFMove(request));
            if (resource->status() != CachedResource::Status::Cached)
                policy = Load;
        } else {
            ResourceError error;
            if (!shouldContinueAfterNotifyingLoadedFromMemoryCache(request,
*resource, error))
```

```cpp
            return makeUnexpected(WTFMove(error));
        logMemoryCacheResourceRequest(frame(),
DiagnosticLoggingKeys::memoryCacheEntryDecisionKey(),
DiagnosticLoggingKeys::usedKey());
        loadTiming.setResponseEnd(MonotonicTime::now());

        memoryCache.resourceAccessed(*resource);

        if (RuntimeEnabledFeatures::sharedFeatures()
            .resourceTimingEnabled() && document() && !resource
            ->isLoading()) {
            auto resourceTiming = ResourceTiming::fromCache(url,
request.initiatorName(), loadTiming, resource->response(), *request.origin());
            if (initiatorContext == InitiatorContext::Worker) {
                ASSERT(is<CachedRawResource>(resource.get()));
                downcast<CachedRawResource>(resource.get())
                  ->finishedTimingForWorkerLoad(WTFMove(resourceTiming));
            } else {
                ASSERT(initiatorContext == InitiatorContext::Document);
                m_resourceTimingInfo
                  .storeResourceTimingInitiatorInformation(resource,
                  request.initiatorName(), frame());
                m_resourceTimingInfo.addResourceTiming(*resource.get(),
*document(), WTFMove(resourceTiming));
            }
        }

        if (forPreload == ForPreload::No)
            resource->setLoadPriority(request.priority());
    }
    break;
}
```

在 CachedResourceLoader 的 loadResource 方法里会通过 Factory 方法 createResource 来创建不同的 CachedResource。

CachedResource 通过加载 CachedResource 实现了 RFC2616 的 W3C 缓存标准。根据资源类型派生出不同的子类，比如 CachedImage、CachedScript、CachedCSSStyleSheet、CachedFont 和 CachedXSLStyleSheet 等。

WebKit 产生一个 CachedResourceClient 的集合 m_clients 用来记录所有的 CachedResource。同时 CachedResourceClientWalker 这个接口可以遍历这个集合。CachedResource 可以通过 CachedResourceClient::notifyFinished 接口等待数据到位，待数据到位后进行通知。

在 CachedResource 里 type 枚举定义了派生资源的类型。通过 defaultPriorityForResourceType 函数方法的返回结果，可以看到各资源加载的优先级，具体定义如下：

```cpp
ResourceLoadPriority CachedResource::defaultPriorityForResourceType(
  Type type)
{
```

```cpp
    switch (type) {
    case CachedResource::MainResource:
        return ResourceLoadPriority::VeryHigh;
    case CachedResource::CSSStyleSheet:
    case CachedResource::Script:
        return ResourceLoadPriority::High;
#if ENABLE(SVG_FONTS)
    case CachedResource::SVGFontResource:
#endif
    case CachedResource::MediaResource:
    case CachedResource::FontResource:
    case CachedResource::RawResource:
    case CachedResource::Icon:
        return ResourceLoadPriority::Medium;
    case CachedResource::ImageResource:
        return ResourceLoadPriority::Low;
#if ENABLE(XSLT)
    case CachedResource::XSLStyleSheet:
        return ResourceLoadPriority::High;
#endif
    case CachedResource::SVGDocumentResource:
        return ResourceLoadPriority::Low;
    case CachedResource::Beacon:
        return ResourceLoadPriority::VeryLow;
#if ENABLE(LINK_PREFETCH)
    case CachedResource::LinkPrefetch:
        return ResourceLoadPriority::VeryLow;
    case CachedResource::LinkSubresource:
        return ResourceLoadPriority::VeryLow;
#endif
#if ENABLE(VIDEO_TRACK)
    case CachedResource::TextTrackResource:
        return ResourceLoadPriority::Low;
#endif
    }
    ASSERT_NOT_REACHED();
    return ResourceLoadPriority::Low;
}
```

WebResourceLoadScheduler 类的 scheduleLoad 方法通过上面返回的优先级来安排整个加载的过程。当然这些实现和平台是相关的，它们会根据不同平台做不同的处理，具体处理如下：

```cpp
void WebResourceLoadScheduler::scheduleLoad(ResourceLoader*
  resourceLoader)
{
    ASSERT(resourceLoader);

#if PLATFORM(IOS)
    // If there's a web archive resource for this URL,
    // we don't need to schedule the load
    // since it will never touch the network.
    if (!isSuspendingPendingRequests() && resourceLoader
```

```cpp
      ->documentLoader()->archiveResourceForURL(resourceLoader
      ->iOSOriginalRequest().url())) {
        resourceLoader->startLoading();
        return;
    }
#else
    if (resourceLoader->documentLoader()
      ->archiveResourceForURL(resourceLoader->request().url())) {
        resourceLoader->start();
        return;
    }
#endif

#if PLATFORM(IOS)
    HostInformation* host = hostForURL(resourceLoader
      ->iOSOriginalRequest().url(), CreateIfNotFound);
#else
    HostInformation* host = hostForURL(resourceLoader->url(), CreateIfNotFound);
#endif

    ResourceLoadPriority priority = resourceLoader->request().priority();

    bool hadRequests = host->hasRequests();
    host->schedule(resourceLoader, priority);

#if PLATFORM(COCOA) || USE(CFURLCONNECTION)
    if (ResourceRequest::resourcePrioritiesEnabled()
      && !isSuspendingPendingRequests()) {
        // Serve all requests at once to keep the pipeline full
        // at the network layer.
        // FIXME: Does this code do anything useful,
        // given that we also set maxRequestsInFlightPerHost to
        // effectively unlimited on these platforms?
        servePendingRequests(host, ResourceLoadPriority::VeryLow);
        return;
    }
#endif

#if PLATFORM(IOS)
    if ((priority > ResourceLoadPriority::Low || !resourceLoader
      ->iOSOriginalRequest().url().protocolIsInHTTPFamily()
      || (priority == ResourceLoadPriority::Low && !hadRequests))
      && !isSuspendingPendingRequests()) {
        servePendingRequests(host, priority);
        return;
    }
#else
    if (priority > ResourceLoadPriority::Low || !resourceLoader->url()
      .protocolIsInHTTPFamily() || (priority == ResourceLoadPriority
      ::Low && !hadRequests)) {
        servePendingRequests(host, priority);
        return;
    }
#endif
```

```
    // Handle asynchronously so early low priority requests don't
    // get scheduled before later high priority ones.
    scheduleServePendingRequests();
}
```

servePendingRequests 有两个重载实现，第一个会根据 host 来遍历，然后由第二个重载来具体实现，具体代码如下：

```
void WebResourceLoadScheduler::servePendingRequests(HostInformation* host,
ResourceLoadPriority minimumPriority)
{
    auto priority = ResourceLoadPriority::Highest;
    while (true) {
        auto& requestsPending = host->requestsPending(priority);
        while (!requestsPending.isEmpty()) {
            RefPtr<ResourceLoader> resourceLoader = requestsPending
                                                        .first();

            // For named hosts - which are only http(s) hosts -
            // we should always enforce the connection limit.
            // For non-named hosts - everything but http(s) -
            // we should only enforce the limit
            // if the document isn't done parsing
            // and we don't know all stylesheets yet.
            Document* document = resourceLoader->frameLoader() ?
resourceLoader->frameLoader()->frame().document() : 0;
            bool shouldLimitRequests = !host->name().isNull() || (document &&
(document->parsing() || !document->haveStylesheetsLoaded()));
            if (shouldLimitRequests && host->limitRequests(priority))
                return;

            requestsPending.removeFirst();
            host->addLoadInProgress(resourceLoader.get());
#if PLATFORM(IOS)
            if (!IOSApplication::isWebProcess()) {
                resourceLoader->startLoading();
                return;
            }
#endif
            resourceLoader->start();
        }
        if (priority == minimumPriority)
            return;
        --priority;
    }
}
```

加载完数据后，SubresourceLoader 处理数据的过程和 MainResource 类似，SubresourceLoader 会作为 ResourceHandle 的接受者，接受 ResourceHandle 的回调。和 MainResource 不同的只是 MemoryCache 部分，这部分的实现在 SubresourceLoader 的 didReceiveDataOrBuffer 里。实现代码如下：

```
    void SubresourceLoader::didReceiveDataOrBuffer(const char* data, int length,
RefPtr<SharedBuffer>&& buffer, long long encodedDataLength, DataPayloadType
dataPayloadType)
    {
        ASSERT(m_resource);

        if (m_resource->response().httpStatusCode() >= 400
&& !m_resource->shouldIgnoreHTTPStatusCodeErrors())
            return;
        ASSERT(!m_resource->resourceToRevalidate());
        ASSERT(!m_resource->errorOccurred());
        ASSERT(m_state == Initialized);
        // Reference the object in this method since
        // the additional processing can do
        // anything including removing the last reference to this object;
        // one example of this is 3266216.
        Ref<SubresourceLoader> protectedThis(*this);

        ResourceLoader::didReceiveDataOrBuffer(data, length, buffer.copyRef(),
encodedDataLength, dataPayloadType);

        if (!m_loadingMultipartContent) {
            if (auto* resourceData = this->resourceData())
                m_resource->addDataBuffer(*resourceData);
            else
                m_resource->addData(buffer ? buffer->data() : data, buffer ?
buffer->size() : length);
        }
    }
```

以 image 加载的顺序为例,DocLoader 负责子资源的加载,例如加载一个图像。首先 DocLoader 要检查 Cache 中是否有对应的 CachedImage 对象。如果有,就不用再次下载和解码,直接用即可,不会再创建一个新的 CachedImage 对象。这个 CachedImage 会要求 Loader 对象创建 SubresourceLoader 来启动网络请求。

关于 CachedImage 的加载过程,具体调用方法的顺序如下:

```
HTMLImageElement::create
ImageLoader::updateFromElementIgnoringPreviousError
ImageLoader::updateFromElement
CachedResourceLoader::requestImage
CachedResourceLoader::requestResource
CachedResourceLoader::loadResource
MemoryCache::add
CachedImage::load
CachedResource::load
CachedResourceLoader::load
CachedResourceRequest::load
ResourceLoaderScheduler::scheduleSubresourceLoad
SubresourceLoader::create
ResourceLoadScheduler::requestTimerFired
ResourceLoader::start
ResourceHandle::create
ResourceLoader::didReceiveResponse
```

```
SubresourceLoader::didiReceiveResponse
CachedResourceRequest::didReceiveResponse
ResourceLoader::didReceiveResponse
ResourceLoader::didReceiveData
SubresourceLoader::didReceiveData
ResourceLoader::didReceiveData
ResourceLoader::addData
CachedResourceRequest::didReceiveData
ResourceLoader::didFinishLoading
SubresourceLoader::didFinishLoading
CachedResourceRequest::didFinishLoading
CachedResource::finish
CachedResourceLoader::loadDone
CachedImage::data
```

一般的资源加载是异步执行的,但是当 JavaScript 文件加载时会阻碍主线程的渲染。这时 WebKit 会在文件加载遇到阻碍时去收集其他资源的 URL,然后通过收集到的资源 URL 来发起并发送请求,这样对加载速度会有一定的优化。

资源的生命周期

资源池采用的是最少使用(Least Recent Used,LRU)算法。通过 HTTP 协议中的更新策略确定下次是否需要更新资源。具体做法是先判断资源是否在资源池里,如果在,就发一个 HTTP 请求,带上一些信息,比如资源的修改时间等。然后服务器根据这个信息判断是否有更新,如果没有更新,就返回 304 状态码。这样就直接使用资源池的原资源,不然就下载最新资源。

WebKit 的 Cache

WebKit 主要有三种 Cache,包括 Page Cache、Memory Cache 和 Disk Cache。

先介绍一下 Page Cache。Page Cache 用于缓存浏览过的页面,这样在用户浏览历史页面时,加载速度会比没有缓存的情况要快。单例模式会有页面上限,会缓存 DOM Tree 和 Render Tree。在打开一个新页面的时候会在 FrameLoader 的 commitProvisionalLoad 方法里,把前一个页面加入 Page Cache。相关代码如下:

```
if (!m_frame.tree().parent() && history().currentItem()) {
    // Check to see if we need to cache the page
    // we are navigating away from into the back/forward cache.
    // We are doing this here because we know for sure that
    // a new page is about to be loaded.
    PageCache::singleton().addIfCacheable(*history().currentItem(),
m_frame.page());

    WebCore::jettisonExpensiveObjectsOnTopLevelNavigation();
}
```

Page Cache 的 addIfCacheable 会对缓存进行管理,实现代码如下:

```
void PageCache::addIfCacheable(HistoryItem& item, Page* page)
{
    if (item.isInPageCache())
```

```cpp
        return;

    if (!page || !canCache(*page))
        return;

    ASSERT_WITH_MESSAGE(!page->isUtilityPage(), "Utility pages such as SVGImage
pages should never go into PageCache");

    setPageCacheState(*page, Document::AboutToEnterPageCache);

    // Focus the main frame, defocusing a focused subframe (if we have one).
    // We do this here,
    // before the page enters the page cache,
    // while we still can dispatch DOM blur/focus events.
    if (page->focusController().focusedFrame())
        page->focusController().setFocusedFrame(&page->mainFrame());

    // Fire the pagehide event in all frames.
    firePageHideEventRecursively(page->mainFrame());

    // Check that the page is still page-cacheable after
    // firing the pagehide event. The JS event handlers
    // could have altered the page in a way that could prevent caching.
    if (!canCache(*page)) {
        setPageCacheState(*page, Document::NotInPageCache);
        return;
    }

    destroyRenderTree(page->mainFrame());

    setPageCacheState(*page, Document::InPageCache);

    // Make sure we no longer fire any JS events past this point.
    NoEventDispatchAssertion assertNoEventDispatch;
    item.m_cachedPage = std::make_unique<CachedPage>(*page);
    item.m_pruningReason = PruningReason::None;
    m_items.add(&item);

    prune(PruningReason::ReachedMaxSize);
}
```

当调用 Page 的 goBack 方法时，会调用 FrameLoader 的 loadDifferentDocumentItem，同时加载类型设置为 FrameLoadTypeBack，然后页面将会从 PageCache 里恢复。

MemoryCache 使得相同 URL 的资源能够快速被获得。单例模式，在浏览器打开时创建，关闭时释放。

MemoryCache 有三个比较重要的属性。

- m_resources：类型是 HashMap，key 是 URL，值是 CacheResource。
- m_allResources：采用的是 LRU 算法，类型是 Vector<LRUList,32>，有 32 个向量，具体实现在 MemoryCache 的 lruListFor 里。回收资源则是通过 MemoryCache 里的 prune，从链表的结尾开始进行回收，直到空间大小达标为止。

- m_liveDecodedResource：类型为 LRUList，解码后的数据会记录在里面。

最后一种 Disk Cache，根据 HTTP 的头信息来设置缓存。属于持久化缓存存储，重新打开浏览器还能节省下载的流量和时间。Disk Cache 使用 LRU 算法来控制缓存磁盘的空间大小。具体实现依据不同平台，实现的代码也不一样，但都是用的 LRU 算法。

3.4.5　HTML 词法解析

解析成 HTML Token 的算法

将 HTML 代码解析成 HTML Token 的算法示意图如下：

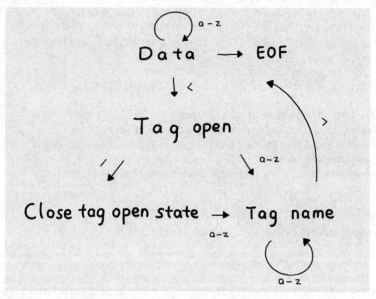

初始解析状态是 Data，碰到 < 符号时，解析状态就会变为 Tag open，下一个字符为字母时就创建 Start tag token，解析状态就变成 Tag name，碰到 > 符号时，解析状态就变成 Data，中间的字符都会被加到 token name 里。

当解析状态是 Tag open 时，如果碰到 / 符号，就会创建 End tag token，同时解析状态会变成 Tag name。在 WebKit 中有两个类是专门做词法分析的，一个是做标记的 HTML Token，另一个是词法解析器 HTML Tokenizer 类，解析任务就是要把字节流解析成一个个 HTML Token，然后给语法解析器分析。在 HTML 解析中一个 Element 对应三个 HTML Token，一个是起始标签，一个是 Element 内容，一个是结束标签。在 DOM Tree 上起始标签和结束标签是相同的 Element Node，Element 内容的 Element Node 和起始、结束标签的 Element Node 不一样。

HTML Token 的所有类型定义在 HTMLToken.h 里。

```
enum Type {
    Uninitialized,    //默认类型
    DOCTYPE,          //文档类型
    StartTag,         //起始标签
```

```
    EndTag,         //结束标签
    Comment,        //注释
    Character,      //元素内容
    EndOfFile,      //文档结束
};
```

接下来看一看 HTML Token 的成员变量,HTML Tokenizer 类会解析出以下结构体。

```
private:
    Type m_type;        //类型

    DataVector m_data;  //不同类型的内容不一样
    UChar m_data8BitCheck;

    // For StartTag and EndTag
    bool m_selfClosing;                 //是否是自封闭
    AttributeList m_attributes;         //属性列表
    Attribute* m_currentAttribute;      //当前属性

    // For DOCTYPE
    std::unique_ptr<DoctypeData> m_doctypeData;

    unsigned m_attributeBaseOffset { 0 }; // Changes across document.write() boundaries.
};
```

HTML Tokenizer 类的 processToken 方法具体实现了整个状态机,主要是根据 W3C 的 tokenization 标准来实现的 。HTML 的字符集合就是数学模型里的字母表,状态的非空集合都定义在 HTMLTokenizer.h 的 State 枚举里。State 枚举值 DataState 表示初始状态。processToken 方法里对不同 State 所做的 case 处理就是状态转移函数。下面是 HTML Tokenizer 类的 State 枚举:

```
enum State {
    DataState,
    CharacterReferenceInDataState,
    RCDATAState,
    CharacterReferenceInRCDATAState,
    RAWTEXTState,
    ScriptDataState,
    PLAINTEXTState,
    TagOpenState,
    EndTagOpenState,
    TagNameState,

    RCDATALessThanSignState,
    RCDATAEndTagOpenState,
    RCDATAEndTagNameState,

    RAWTEXTLessThanSignState,
    RAWTEXTEndTagOpenState,
    RAWTEXTEndTagNameState,

    ScriptDataLessThanSignState,
```

```
ScriptDataEndTagOpenState,
ScriptDataEndTagNameState,
ScriptDataEscapeStartState,
ScriptDataEscapeStartDashState,
ScriptDataEscapedState,
ScriptDataEscapedDashState,
ScriptDataEscapedDashDashState,
ScriptDataEscapedLessThanSignState,
ScriptDataEscapedEndTagOpenState,
ScriptDataEscapedEndTagNameState,
ScriptDataDoubleEscapeStartState,
ScriptDataDoubleEscapedState,
ScriptDataDoubleEscapedDashState,
ScriptDataDoubleEscapedDashDashState,
ScriptDataDoubleEscapedLessThanSignState,
ScriptDataDoubleEscapeEndState,

BeforeAttributeNameState,
AttributeNameState,
AfterAttributeNameState,
BeforeAttributeValueState,
AttributeValueDoubleQuotedState,
AttributeValueSingleQuotedState,
AttributeValueUnquotedState,
CharacterReferenceInAttributeValueState,
 AfterAttributeValueQuotedState,
SelfClosingStartTagState,
BogusCommentState,
// Not in the HTML spec,
// used internally to track whether we started the bogus comment token.
ContinueBogusCommentState,
MarkupDeclarationOpenState,

CommentStartDashState,
CommentState,
CommentEndDashState,
CommentEndState,
CommentEndBangState,

DOCTYPEState,
BeforeDOCTYPENameState,
DOCTYPENameState,
AfterDOCTYPENameState,
AfterDOCTYPEPublicKeywordState,
BeforeDOCTYPEPublicIdentifierState,
DOCTYPEPublicIdentifierDoubleQuotedState,
DOCTYPEPublicIdentifierSingleQuotedState,
AfterDOCTYPEPublicIdentifierState,
BetweenDOCTYPEPublicAndSystemIdentifiersState,
AfterDOCTYPESystemKeywordState,
BeforeDOCTYPESystemIdentifierState,
DOCTYPESystemIdentifierDoubleQuotedState,
DOCTYPESystemIdentifierSingleQuotedState,
AfterDOCTYPESystemIdentifierState,
```

```
    BogusDOCTYPEState,
    CDATASectionState,

    CDATASectionRightSquareBracketState,
    CDATASectionDoubleRightSquareBracketState,
};
```

词法解析的主要接口是 nextToken 函数。

3.4.6 HTML 语法解析

HTML 的语法定义

HTML 的语法规范是由 W3C 组织创建和定义的。语法、语句可以使用正式的范式，比如使用 BNF 范式来定义 HTML 语法规范，不过对于 HTML 这种会在创建后允许再插入代码的语言，W3C 使用 DTD 来定义 HTML。

主要设计类

- Document 和 DocumentParser 相互引用。
- HTMLDocumentParser 的父类是 ScriptableDocumentParser，ScriptableDocumentParser 的父类是 DecodedDataDocument Parser，DecodedDataDocument Parser 的父类是 Document Parser。HTMLInputStream 解码后会保存字符流，将保存的字符流作为缓冲区。
- HTMLTokenizer：对字符流进行解析，将解析后的信息，比如 Tag 名属性列表注释、Doc 声明、结束 Tag、文本都存在 HTMLToken 类中。
- HTMLTreeBuilder：生成 DOM Tree，这里会做语义上的检查，比如 head tag 里不能包含 body tag 等。
- HTMLToken：HTMLToken 是由 HTMLTokenizer 生成的，然后 HTMLTree Builder 会使用 HTMLToken 来构建 DOM Tree。
- HTMLConstructionSite：创建 HTMLElement，再完成 DOM Tree，关键标签除 body、head 和 html 以外，都是通过 HTMLElementFactory 来实现的。其属性是由生成的 Element 里对应的函数来实现的。
- HTMLPreloadScanner：专门用来查找类似 src、link 这样的属性，获取外部资源，通过 CachedResourceLoader 类进行加载。

解析过程

完整解析过程的示意图如下：

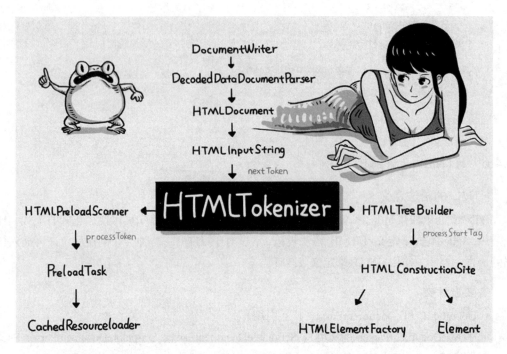

- DocumentWriter 调用 DocumentParser::appendBytes，将结果存入 HTMLInputStream 中。
- HTMLDocumentParser::pumpTokenizer 函数里的 nextToken 会把字符流解析成 HTMLToken。
- 把 HTMLToken 对象交给 HTMLTreeBuilder，通过 HTMLTreeBuilder:: processToken 分析标签是否正确，再将 HTMLConstructionSite 的 insertHTMLElement、insertHTMLHeadElement 插入到 HTMLConstructionSite 里。HTMLConstructionSite 通过 HTMLElementFactory 里的 createHTMLElement 来创建 HTMLElement 对象。Element 的 parserSetAttributes 函数用来解析属性。创建完的 Element 则会被添加到当前的 Node 里，通过 HTMLConstructionSite 的 attachLater 函数异步完成。
- 然后 HTMLToken 对象还会传给 HTMLPreloadScanner，它会先调用 scan 函数，生成一个 PreloadTask，再调用 preload 函数或 CachedResourceLoader 的 preload 进行资源预加载。CachedResourceLoader 里的资源是以 URL 作为 key 的，通过 URL 来获取资源。具体实现可以看 HTMLPreloadScanner::processToken。

Element 属性设置

Element::parserSetAttributes 的设置如下所示。

```
void Element::parserSetAttributes(const Vector<Attribute>& attributeVector)
{
    ASSERT(!isConnected());
    ASSERT(!parentNode());
    ASSERT(!m_elementData);

    if (!attributeVector.isEmpty()) {
        if (document().sharedObjectPool())
```

```
            m_elementData = document().sharedObjectPool()->
                cachedShareableElementDataWithAttributes(attributeVector);
        else
            m_elementData = ShareableElementData::createWithAttributes(
                attributeVector);

    }

    parserDidSetAttributes();

    // Use attributeVector instead of m_elementData
    // because attributeChanged might modify m_elementData.
    for (const auto& attribute : attributeVector)
        attributeChanged(attribute.name(), nullAtom(), attribute.value(),
ModifiedDirectly);
    }
```

m_elementData 是存储属性的对象。attributeChanged 函数会调用 parseAttribute 函数，作用是让 Element 能够解析属性，比如图片标签中用来加载图片的 src 属性。

资源加载

HTMLPreloadScanner 通过调用 CachedResourceLoader 来实现加载。

3.4.7　构建 DOM Tree

构建 DOM Tree 需要经历解码（decode）、分词（tokenizer）、解析（Parse，创建与 Tag 相对应的 Node）、创建树（Attach，可创建三种树，DOM Tree、Render Tree 和 RenderLayer Tree）这四个阶段，如下图所示。

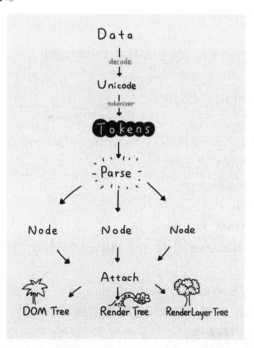

构建 DOM Tree 的四个阶段，如下图所示。

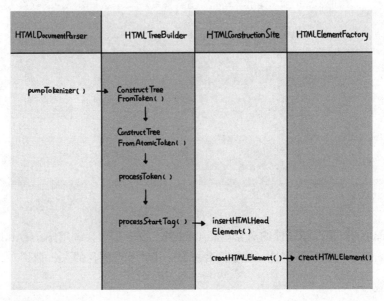

DOM 标准规范

DOM（Document Object Model，文档对象模型）可以支持 HTML、XML 和 XHTML 三种文件类型。DOM 是以面向对象的方式来描述文档的，因此可以通过 JavaScript 或其他面向对象的语言访问、创建、删除和修改 DOM 结构。DOM 的接口与平台和语言无关。W3C 定义了一系列的 DOM 接口，目前已经有了 3 个 level 标准，每个标准都是在上一个标准的基础上进行迭代的。

下面对各个 level 进行说明：

DOM level 1
- Core：底层接口，接口支持 XML 等任何结构化文档。
- HTML：把 HTML 内容定义为 Document（文档）、Node（节点）、Attribute（属性）、Element（元素）和 Text（文本）。

DOM level 2
- Core：对 level 1 的 Core 部分进行扩展，比如 getElementById 和 namespace 等接口。
- HTML：允许动态访问修改文档。
- Views：文档的各种视图。
- Events：鼠标事件等。
- Style：可以修改 HTML 样式的一个属性。
- Traveral and range：NodeIterator 和 TreeWalker 遍历树对文档进行修改、删除等操作。

DOM level 3
- Core：加入了新接口 adoptNode 和 textContent。
- Load and Save：加载 XML，转成 DOM 表示的文档结构。
- Validation：验证文档有效性。

- Events：加入键盘支持。
- XPath：一种简单、直观的检索 DOM 节点的方式。

DOM 结构构成的基本要素是节点，节点概念比较大，Document 也是一个节点，叫 Document 节点；HTML 里的 Tag 也是一种节点，叫 Element 节点。还有 Attribute 节点、Entity 节点、ProcessingInstruction 节点、CDataSection 节点、Comment 节点。DOM 节点接口的描述文件路径在 WebCore/dom/Document.idl 文件里。DOM Tree Token 的构造算法被描述为一个状态机，State 为插入模式。

当 Parser 创建后，Parser 会先创建一个 Document 对象。在 tree construction 的阶段，Document 会作为 root 的 Element 添加进来。tree constructor 会处理每个 tokenizer 生成的 Node。每个 token 会对应一个 DOM Element。每个 Element 不仅会被添加到 DOM Tree，还会添加到 stack of open elements 里。这个 stack 用于检查嵌套错误和未关闭的 Tag。

具体过程是，在 tree construction 阶段的输入来自 tokenization 对 token 排序后的结果。最开始的状态是 initial mode。接收 html token 会变成 before html mode，并且在 before html mode 里重新处理 token。这会创建一个 HTMLHtmlElement 元素并添加到 Document 对象的根节点。

接下来状态会变成 before head，然后创建 HTMLHeadElement。如果 head 里有 token，就会被添加到这个 Tree 里。状态有 in head 和 after head。

当状态为 after head 时会创建 HTMLBodyElement，添加到 Tree 中，状态转成 in body，其中 html token 会被转成对应的 DOM Element，添加到 body 的 Node 中。当收到 body end token 时会让状态切换成 after body。接着就收到 html end tag，转成 after after body 状态，直到 end of file token 结束解析。

关键类

构建 DOM 的几个关键的类，如下所示。

- HTMLDocumentParser：管理类，解析 HTML 文本，使词法分析器成为 HTMLTokenizer 类，最后输出到 HTMLTreeBuilder。
- HTMLTreeBuilder：负责 DOM Tree 的建立，对 token 分类处理，创建一个个节点对象，使用 HTMLConstructionSite 类将这些创建好的节点构建为 DOM Tree。
- HTMLConstructionSite：不同的标签类型在不同位置会有不同的函数来构建 DOM Tree。m_document 代表根节点，即 JavaScript 里的 window.document 对象。
- HTMLElementFactory：一个 Factory 类对不同类型的标签创建不同的 HTML 元素，同时建立父子关系或兄弟关系。

举一个例子，代码如下：

```
<html>
  <body>
    <p>
      This is a paragraph
    </p>
    <div><p>new paragraph</p></div>
```

```
        </body>
</html>
```

将其转成 DOM Tree 的过程示意图如下：

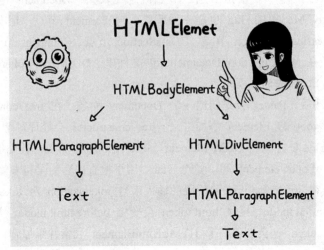

加载

通过 DocumentLoader 类的 startLoadingMainResource 加载 URL，代码如下：

```
FetchRequest fetchRequest(m_request, FetchInitiatorTypeNames::document,
                    mainResourceLoadOptions);
  m_mainResource =
      RawResource::fetchMainResource(fetchRequest, fetcher(), m_substituteData);
// DocumentLoader 的 commitData 会处理 dataReceived 的数据块
void DocumentLoader::commitData(const char* bytes, size_t length) {
  //初始化 HTMLDocumentParser，实例化 Document 对象
  ensureWriter(m_response.mimeType());
  if (length)
    m_dataReceived = true;
  //给 Parser 解析，这里的 bytes 就是返回来的 HTML 文本代码
  m_writer->addData(bytes, length);
}
//ensureWriter 里有个判断，如果 m_writer 已经初始化就不处理
//所以 Parser 只会初始化一次
    void DocumentLoader::ensureWriter(const AtomicString& mimeType, const KURL&
overridingURL) {
  if (m_writer)
    return;
}
```

DOM Tree

DOM Tree 已经被 W3C 标准化了。Document Object Model (DOM) Technical Reports 在 DOM Level 3 里，IDL 的定义在 IDL Definitions 里。

Node 是 DOM 模型的基础类，根据标签的不同可以分成多个类，详细的定义可以看 NodeType。其中最主要的是 Document、Element 和 Text 三个类。下图列出了 Node 的继承关系图：

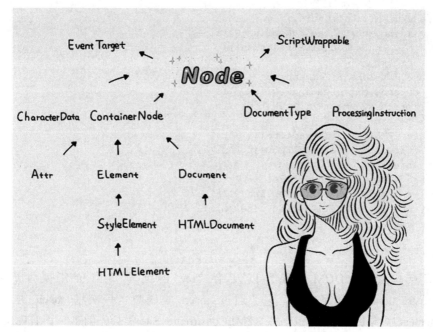

在构建树时,由 Insertion Mode 来控制。这个规则是完全按照 W3C 指定的,根据 token 的堆栈来构建树级别,碰到 StartTag 的 token 就会在 m_openElements 里压栈,碰到 EndTag 的 token 就会出栈。W3C 对于 Element 的堆栈也有规范。通过 ContainerNode::parserAddChild 接口,将 Node 添加到 DOM Tree 中。

需要注意的是,有些标签属于 Format 标签,Format 标签属于平级关系,不参与嵌套关系,不会被加入 m_openElements 里。

tag 对应的 Node 是由 HTMLElementFactory 类中的 createHTMLElement 函数来创建的,然后通过 Hash Map 来记录,key 为 tag name,value 为对应的 Element。不同 tag 有不同的 HTMLElement 派生类。

构建 DOM Tree 是从 token 构建节点开始的,在 HTMLDocumentParser 类中通过调用 HTMLTreeBuilder 类的 processToken 函数来完成 token 构建节点的过程。processToken 函数在 WebCore/html/parser/HTMLTreeBuilder.cpp 文件里,代码如下:

```
void HTMLTreeBuilder::processToken(AtomicHTMLToken&& token)
{
    switch (token.type()) {
    case HTMLToken::Uninitialized:
        ASSERT_NOT_REACHED();
        break;
    case HTMLToken::DOCTYPE:
        m_shouldSkipLeadingNewline = false;
        processDoctypeToken(WTFMove(token));
        break;
    case HTMLToken::StartTag:
        m_shouldSkipLeadingNewline = false;
        processStartTag(WTFMove(token));
        break;
```

```
        case HTMLToken::EndTag:
            m_shouldSkipLeadingNewline = false;
            processEndTag(WTFMove(token));
            break;
        case HTMLToken::Comment:
            m_shouldSkipLeadingNewline = false;
            processComment(WTFMove(token));
            return;
        case HTMLToken::Character:
            processCharacter(WTFMove(token));
            break;
        case HTMLToken::EndOfFile:
            m_shouldSkipLeadingNewline = false;
            processEndOfFile(WTFMove(token));
            break;
    }
}
```

将节点构建成 DOM Tree，以及给 Tree 的元素节点创建属性，都是由 HTMLConstructionSite 类完成的。它会通过 insert 类函数，一次将 token 类型创建的 HTMLElement 对象加入 DOM Tree。HTMLConstructionSite 类会先创建一个 HTML 根节点 HTMLDocument 对象和一个栈 HTMLElementStack 变量，栈里存放没有遇到 EndTag 的所有 StartTag。根据这个栈来建立父子关系，代码如下：

```
    void HTMLConstructionSite::insertHTMLHtmlStartTagBeforeHTML(
      AtomicHTMLToken* token) {
      HTMLHtmlElement* element = HTMLHtmlElement::create(*m_document);
      attachLater(m_attachmentRoot, element);
      m_openElements.pushHTMLHtmlElement(HTMLStackItem::create(element, token));
//push 到 HTMLStackItem 栈里
      executeQueuedTasks();
    }
//通过 attachLater 创建 task
    void HTMLConstructionSite::attachLater(ContainerNode* parent,
                                            Node* child,
                                            bool selfClosing) {
      HTMLConstructionSiteTask task(HTMLConstructionSiteTask::Insert);
      task.parent = parent;
      task.child = child;
      task.selfClosing = selfClosing;
      //判断是否达到最深，512 是最深的
      if (m_openElements.stackDepth() > maximumHTMLParserDOMTreeDepth &&
          task.parent->parentNode())
        task.parent = task.parent->parentNode();
      queueTask(task);
    }
//executeQueued 添加子节点
    void ContainerNode::parserAppendChild(Node* newChild) {
      if (!checkParserAcceptChild(*newChild))
        return;
      AdoptAndAppendChild()(*this, *newChild, nullptr);
    }
```

```
  notifyNodeInserted(*newChild, ChildrenChangeSourceParser);
}
//添加前先检查是否支持子元素
void ContainerNode::appendChildCommon(Node& child) {
  child.setParentOrShadowHostNode(this);
  if (m_lastChild) {
    child.setPreviousSibling(m_lastChild);
    m_lastChild->setNextSibling(&child);
  } else {
    setFirstChild(&child);
  }
  setLastChild(&child);
}
//是 EndTag 就会对元素进行出栈操作
m_tree.openElements()->popUntilPopped(token->name());
```

对错误的处理

解析一个 HTML 代码时,如果遇到不符合规范的代码,就需要 Parser 做容错。下面列出了一些 WebKit 容错的例子。例如,br 的问题,一些网站会用 /br 替代 br,处理的代码如下:

```
if (t->isCloseTag(brTag) && m_document->inCompatMode()) {
    reportError(MalformedBRError);
    t->beginTag = true;
}
```

stray table 是指一个 table 包含了一个不在 table cell 的 table,代码如下:

```
<table>
  <table>
    <tr><td>inner table</td></tr>
      </table>
  <tr><td>outer table</td></tr>
</table>
```

WebKit 的处理代码如下:

```
if (m_inStrayTableContent && localName == tableTag)
    popBlock(tableTag);
```

这样会处理成两个同级 table,代码如下:

```
<table>
  <tr><td>outer table</td></tr>
</table>
<table>
  <tr><td>inner table</td></tr>
</table>
```

关于嵌套 form 的问题,将一个 form 放到另一个 form 里。那么第二个 form 会被忽略,代码如下:

```
if (!m_currentFormElement) {
    m_currentFormElement = new HTMLFormElement(formTag,   m_document);
}
```

关于过深的层级问题，同一个类型里只允许嵌套 20 个 Tag，代码如下：

```
bool HTMLParser::allowNestedRedundantTag(const AtomicString& tagName)
{
unsigned i = 0;
for (HTMLStackElem* curr = m_blockStack;
     i < cMaxRedundantTagDepth && curr && curr->tagName == tagName;
   curr = curr->next, i++) { }
return i != cMaxRedundantTagDepth;
}
```

html 或 body 的 end tags 缺失都会在 end() 里调用，代码如下：

```
if (t->tagName == htmlTag || t->tagName == bodyTag )
       return;
```

3.4.8 CSS

CSS 语法简介

所有的 CSS 都是由 CSSStyleSheet 集合组成的。CSSStyleSheet 是由多个 CSSRule 组成的。CSSRule 是由 CSSStyleSelector 选择器和 CSSStyleDeclaration 声明组成的。CSSStyleDeclaration 是 CSS 属性的键值集合。

在 CSS Declaration block 前加上 selector 来匹配页面上的元素。selector 和 Declaration block 在一起被称作 "ruleset"，如下图所示。

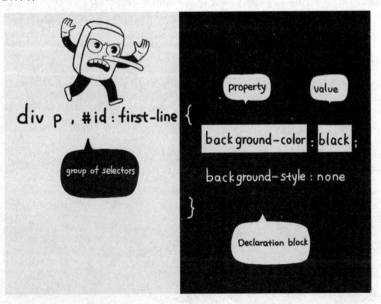

selector 可以分为以下类别。

- Simple selector：简单选择器，通过 class 或 id 类匹配一个或多个元素。
- Attribute selector：属性选择器，通过属性匹配。
- Pseudo-class：伪类，匹配元素的确定状态，比如元素处于鼠标悬停状态；复选框是选中或未选中的状态；如果 DOM Tree 中一个父节点的第一个子节点为 a 标签元素，那么 a:visited 表示访问过的状态， a:hover 表示鼠标悬停在上面。
- Pseudo-element：伪元素，匹配处于确定位置的一个或多个元素，比如每个段落的第一个字、某个元素之前生成的内容。比如 [href^=http]::after { content: '->'} 表示属性是以 herf 值为开头的，并且在 http 的元素内容后面添加 -> 符号。
- Combinator：组合器，指组合多个选择器用于特定的选择方法。比如只选择 div 的直系子节点段落。比如 A > B， B 是 A 的直接子节点。
- Multiple selector：多用选择器，将一组声明应用于由多个选择器选择的所有元素。

Attribute selector：属性选择器是针对 data-* 这样的数据属性进行选择的，尝试匹配精确属性值的 Presence and value attribute selector 分为下面几种类型：

- [attr]：不论值是什么，选择所有 attr 属性的元素。
- [attr=val]：匹配出 attr 属性的值为 val 的所有元素。
- [attr~=val]：匹配出 attr 属性的值包含 val 值的所有元素，val 以空格间隔多个值。

还有 Substring value attribute selector 这样的伪正则选择器：

- [attr|=val]：匹配选择以 val 或 val- 开头的元素，一般用来处理语言编码。
- [attr^=val]：匹配选择以 val 开头的元素。
- [attr$=val]：匹配选择以 val 结尾的元素。
- [attr*=val]：匹配选择包含 val 的元素。

Combinator 组合器有以下几种类型：

- A,B： 匹配 A 、B 两个中的任意元素。
- A B：B 是 A 的后代节点，可以是直接的子节点或者是子节点的子节点。
- A > B：B 是 A 的直接子节点。
- A + B：B 是 A 的兄弟节点。
- A ~ B： B 是 A 兄弟节点中的任意一个。

CSS Rule 只是 CSS statement 的一种。 其他类型是 @-规则，在 CSS 里使用 @-规则来传递元数据、条件信息或其他描述信息。@ 符号后面跟一个描述符来表明用哪种规则，在后面接 CSS declarations block，最后以 ; 符号结束。

- @charset：表示字符集的元数据。
- @import：表示引用的元数据。
- @media：嵌套语句，当运行浏览器的硬件设备匹配了@media 里的条件时，才会运用 @-规则里的内容。
- @supports：嵌套语句，只有浏览器确实支持，才会应用 @-规则里的内容。
- @document：嵌套语句，当前页面匹配了@document 条件时，才会运用 @-规则里的内容。
- @font-face：描述性信息。

向当前的 CSS 中导入 starming.css 文件，如下所示。

```
@import 'starming.css';
```

页面宽度超过 801 像素，从 802 开始时才会运用 @-规则里的内容，如下所示。

```
@media (min-width: 801px) {
    body {
        margin:0 auto;
        width:800px;
    }
}
```

有些属性是 Shorthand 属性，比如 font、background、padding、border 和 margin。它们允许一行里设置多个属性。

```
margin: 20px 10px 10px 20px;
```

等效于：

```
margin-top: 20px;
margin-right: 10px;
margin-bottom: 10px;
margin-left: 20px;
```

再比如：

```
background: black url(starming-bg.png) 20px 20px repeat-x fixed;
```

等效于：

```
background-color: black;
background-image: url(starming-bg.png);
background-position: 20px 20px;
background-repeat: repeat-x;
background-scroll: fixed;
```

- 绝对单位：px、mm、cm、in、pt、pc。
- 相对单位：em。1em 的计算值默认为 16px。但是要注意 em 单位会继承父元素的字号大小。WebCore 对 ex 和 ch 的支持不太好。rem 与 em 的工作方式一样，不同的是，rem 总是等于默认的基础字号大小。vw、vh 分别指视图宽度的 1/100 和视图高度的 1/100，但 WebCore 对 vw 和 vh 的支持并没有对 rem 的支持那样好。
- 无单位的值：在某些情况下，无单位的值是可以存在的，比如设置 margin: 0 即可，或者设置为 p { line-height: 1.5 }，这里的 1.5 类似一个简单的乘法因子，比如字号大小为 16px 的话，行高就为 24px。
- 百分比：表示相对于父容器所占的百分比时，font-size 使用的是百分比；表示相对于父容器的字体大小时，font-size 既可以使用百分比，也可以使用 em。
- 颜色：有 165 个不同关键字可用。还可以使用类似 background-color:#0000ff 这样的十六进制数表示颜色，RGB 表示 background-color:rgb(0,0,255)。WebCore 还支持 background-color:hsl(240,100%,50%) 这种 HSL 的表示，值依次是色调，值范围是 0 到 360，饱和度为 0 表示没有颜色，明度为 0 表示全黑，明度为 100%表示全白。HSL 使用圆柱体可视化表达颜色。
- 透明度：可以通过 rgba 和 hsla 来表示，比如 background-color:hsla(240,100%,50%, 0.5)。

还可以通过 CSS 属性 opacity 来指定透明度。
- 函数：只要是一个名字后面用括号包含一个或多个值，多个值之间用逗号分隔的格式表示的就是一个函数，比如 background-color: rgba(255,0,0,0.5)、transform: translate(50px, 60px)、background-image: url('myimage.png')。

CSS BNF

CSS 有自己的语法 BNF，CSS BNF 描述内容如下：

```
ruleset
  : selector [ ',' S* selector ]*
    '{' S* declaration [ ';' S* declaration ]* '}' S*
  ;
selector
  : simple_selector [ combinator selector | S+ [ combinator selector ] ]
  ;
simple_selector
  : element_name [ HASH | class | attrib | pseudo ]*
  | [ HASH | class | attrib | pseudo ]+
  ;
class
  : '.' IDENT
  ;
element_name
  : IDENT | '*'
  ;
attrib
  : '[' S* IDENT S* [ [ '=' | INCLUDES | DASHMATCH ] S*
    [ IDENT | STRING ] S* ] ']'
  ;
pseudo
  : ':' [ IDENT | FUNCTION S* [IDENT S*] ')' ]
  ;
```

WebKit 使用 Flex 和 Bison 来解析这个 BNF 文件。CSS 文件会被解析成 StyleSheet object，每个 object 包含 CSS Rule object，CSS Rule objcet 包含 selector 和 declaration ，以及其他对应于 CSS 语法的 object。接下来进行 CSSRule 的匹配过程，去找能够和 CSSRule Selector 部分匹配的 HTML 元素。

解析 CSS

WebKit 解析的入口函数是 CSSParser::parseSheet。

将 token 转成 styleRule。每个 styleRule 包含 selector 和 property。

定义 MatchType，代码如下：

```
enum MatchType {
  Unknown,
  Tag,              //比如 div
  Id,               // #id
  Class,            // .class
  PseudoClass,      // :nth-child(2)
```

```
    PseudoElement,      // ::first-line
    PagePseudoClass,    //
    AttributeExact,     // E[attr="value"]
    AttributeSet,       // E[attr]
    AttributeHyphen,    // E[attr|="value"]
    AttributeList,      // E[attr~="value"]
    AttributeContain,   // E[attr*="value"]
    AttributeBegin,     // E[attr^="value"]
    AttributeEnd,       // E[attr$="value"]
    FirstAttributeSelectorMatch = AttributeExact,
};
```

定义 selector 的 Relation 类型，代码如下：

```
enum RelationType {
    SubSelector,        // No combinator
    Descendant,         // "Space" combinator
    Child,              // > combinator
    DirectAdjacent,     // + combinator
    IndirectAdjacent,   // ~ combinator
    // Special cases for shadow DOM related selectors.
    ShadowPiercingDescendant, // >>> combinator
    ShadowDeep,               // /deep/ combinator
    ShadowPseudo,             // ::shadow pseudo element
    ShadowSlot                // ::slotted() pseudo element
};
```

CSS 的属性是用 id 来标识的，代码如下：

```
enum CSSPropertyID {
    CSSPropertyColor = 15,
    CSSPropertyWidth = 316,
    CSSPropertyMarginLeft = 145,
    CSSPropertyMarginRight = 146,
    CSSPropertyMarginTop = 147,
    CSSPropertyMarkerEnd = 148,
}
```

生成 Hash Map，并且分成以下四个类型：

```
CompactRuleMap m_idRules;
CompactRuleMap m_classRules;
CompactRuleMap m_tagRules;
CompactRuleMap m_shadowPseudoElementRules;
```

CSS 解析完会触发 Layout Tree，给每个可视 Node 创建一个 Layout 节点，创建时需要计算 style，这个过程包括找到 selector 和 设置 style。

Layout 会更新递归所有的 DOM 元素，代码如下：

```
void ContainerNode::attachLayoutTree(const AttachContext& context) {
    for (Node* child = firstChild(); child; child = child->nextSibling()) {
        if (child->needsAttach())
            child->attachLayoutTree(childrenContext);
    }
}
```

然后对每个 Node 按照一定顺序取出所有对应的 selector，代码如下：

```
if (element.hasID())
  collectMatchingRulesForList(
      matchRequest.ruleSet->idRules(element.idForStyleResolution()),
      cascadeOrder, matchRequest);
if (element.isStyledElement() && element.hasClass()) {
  for (size_t i = 0; i < element.classNames().size(); ++i)
    collectMatchingRulesForList(
      matchRequest.ruleSet->classRules(element.classNames()[i]),
      cascadeOrder, matchRequest);
}
...
collectMatchingRulesForList(
    matchRequest.ruleSet
      ->tagRules(element.localNameForSelectorMatching()),
    cascadeOrder, matchRequest);
...
```

可以看到 id 是唯一的，可以直接取到，而 class 则需要遍历数组来设置 style。在 classRules 里进行检验，代码如下：

```
if (!checkOne(context, subResult))
  return SelectorFailsLocally;
if (context.selector->isLastInTagHistory()) {
    return SelectorMatches;
}
//checkOne 的实现
switch (selector.match()) {
 case CSSSelector::Tag:
   return matchesTagName(element, selector.tagQName());
 case CSSSelector::Class:
   return element.hasClass() &&
       element.classNames().contains(selector.value());
 case CSSSelector::Id:
   return element.hasID() &&
       element.idForStyleResolution() == selector.value();
}
```

如果 relation 的类型是 Descendant，则表示后代需要在 checkOne 后处理 relation，代码如下：

```
switch (relation) {
 case CSSSelector::Descendant:
   for (nextContext.element = parentElement(context); nextContext
     .element; nextContext.element = parentElement(nextContext)) {
     MatchStatus match = matchSelector(nextContext, result);
     if (match == SelectorMatches || match == SelectorFailsCompletely)
       return match;
     if (nextSelectorExceedsScope(nextContext))
       return SelectorFailsCompletely;
   }
   return SelectorFailsCompletely;
  case CSSSelector::Child:
```

```
    //...
}
```

可以看到这个就是 selector 从右到左解释的实现，它会递归所有父节点，并且再次执行 checkOne，直到找到需要的父节点才会停止递归。所以选择器的节点不要写得太长，否则会影响效率。

设置 style

设置 style 的顺序是先设置父节点，再使用默认的 UA 的 style，最后使用 Author 的 style，代码如下：

```
style->inheritFrom(*state.parentStyle())
matchUARules(collector);
matchAuthorRules(*state.element(), collector);
```

在执行每一步时如有一个 styleRule 匹配，那么 styleRule 会放到当前元素的 m_matchedRule 里，然后计算优先级，优先级算法是从右到左取每个 selector 优先级之和，实现如下：

```
for (const CSSSelector* selector = this; selector;
    selector = selector->tagHistory()) {
  temp = total + selector->specificityForOneSelector();
}
return total;
```

每个不同类型的 selector 的优先级如下：

```
switch (m_match) {
  case Id:
    return 0x010000;
  case PseudoClass:
    return 0x000100;
  case Class:
  case PseudoElement:
  case AttributeExact:
  case AttributeSet:
  case AttributeList:
  case AttributeHyphen:
  case AttributeContain:
  case AttributeBegin:
  case AttributeEnd:
    return 0x000100;
  case Tag:
    return 0x000001;
  case Unknown:
    return 0;
  }
  return 0;
}
```

可以看出 id 的优先级最高为 0x010000 = 65536。class、属性和伪类的优先级是 0x000100 = 256，tag 的优先级是 0x000001 = 1。

举一个优先级计算的例子，代码如下：

```
/*优先级为257 = 256 + 1*/
.text h1{
    font-size: 8em;
}
/*优先级为65537 = 65536 + 1*/
#my-text h1{
    font-size: 16em;
}
```

将所有 CSS 规则数据放到 collector 的 m_matchedRules 数组里，再根据优先级从低到高排序。这样，越执行到后面优先级越高，优先级高的规则可以覆盖前面优先级低的规则。

优先级从高到低的排序是 id 选择器>类型选择器>标签选择器>相邻选择器>子选择器>后代选择器。

关于优先级的排序规则，代码如下：

```
static inline bool compareRules(const MatchedRule& matchedRule1,
                    const MatchedRule& matchedRule2) {
  unsigned specificity1 = matchedRule1.specificity();
  unsigned specificity2 = matchedRule2.specificity();
  if (specificity1 != specificity2)
    return specificity1 < specificity2;

  return matchedRule1.position() < matchedRule2.position();
}
```

规则和优先级确定后，就可以设置元素的 style 了，代码如下：

```
applyMatchedPropertiesAndCustomPropertyAnimations(
      state, collector.matchedResult(), element);

applyMatchedProperties<HighPropertyPriority, CheckNeedsApplyPass>(
    state, matchResult.allRules(), false, applyInheritedOnly,
    needsApplyPass);
  for (auto range : ImportantAuthorRanges(matchResult)) {
    applyMatchedProperties<HighPropertyPriority, CheckNeedsApplyPass>(
      state, range, true, applyInheritedOnly, needsApplyPass);
  }
```

最后的 style 是什么形式？style 会生成一个 ComputedStyle 对象。ComputedStyle 对象结构包括 m_styleInheritedData、m_noneInheritedData、m_svgStyle、m_box、m_background、m_Surround 六大部分，示意图如下：

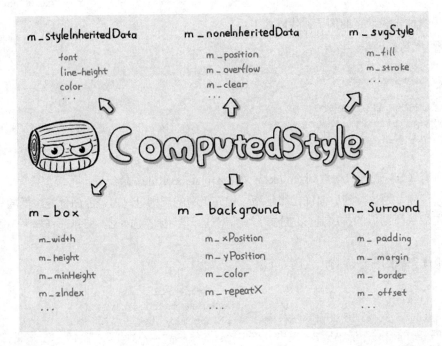

CSSOM

CSSOM 是 CSS 的对象模型,其提供了便于 JavaScript 操作 CSS 的接口 CSSStyleSheet。通过这个接口,我们可以更容易获取 CSS 的 href、CSSRule 等信息。

W3C 还定义了 CSSOM View,增加了 Window、Document、Element、HTMLElement 和 MouseEvent 的接口。比如 Window 对 CSSOM View 的支持就可以在 WebCore/page/DOMWindow.idl 里查看到。其中与 CSSOM View 相关的规范定义如下:

```
    // Extensions from the CSSOM-View specification
(https://drafts.csswg.org/cssom-view/#extensions-to-the-window-interface).
    [NewObject] MediaQueryList matchMedia(CSSOMString query);
    // FIXME: Should be [SameObject].
    [Replaceable] readonly attribute Screen screen;

    // Browsing context (CSSOM-View).
    void moveTo(optional unrestricted float x = NaN, optional unrestricted float y = NaN); // FIXME: Parameters should be mandatory and of type long.
    void moveBy(optional unrestricted float x = NaN, optional unrestricted float y = NaN); // FIXME: Parameters should be mandatory and of type long.
    // Parameters should be mandatory and of type long.
    void resizeTo(optional unrestricted float width = NaN, optional unrestricted float height = NaN);
    void resizeBy(optional unrestricted float x = NaN, optional unrestricted float y = NaN); // FIXME: Parameters should be mandatory and of type long.

    // Viewport (CSSOM-View).
    [Replaceable] readonly attribute long innerHeight;
    [Replaceable] readonly attribute long innerWidth;

    // Viewport scrolling (CSSOM-View).
```

```
[Replaceable] readonly attribute double scrollX;
[Replaceable, ImplementedAs=scrollX] readonly attribute double pageXOffset;
[Replaceable] readonly attribute double scrollY;
[Replaceable, ImplementedAs=scrollY] readonly attribute double pageYOffset;
[ImplementedAs=scrollTo] void scroll(optional ScrollToOptions options);
[ImplementedAs=scrollTo] void scroll(unrestricted double x, unrestricted double y);
void scrollTo(optional ScrollToOptions options);
void scrollTo(unrestricted double x, unrestricted double y);
void scrollBy(optional ScrollToOptions option);
void scrollBy(unrestricted double x, unrestricted double y);

// Client (CSSOM-View).
[Replaceable] readonly attribute long screenX;
[Replaceable] readonly attribute long screenY;
[Replaceable] readonly attribute long outerWidth;
[Replaceable] readonly attribute long outerHeight;
[Replaceable] readonly attribute double devicePixelRatio;
```

可以看出 CSSOM View 定义了 Window 的大小、滚动、位置等参数。

3.4.9　RenderObject Tree

在 DOM Tree 构建过程中，在 HTMLConstruction 的 attachToCurrent 方法里通过 createRendererIfNeeded 方法构建 RenderTree。构成这个 Tree 的元素是 RenderObject。RenderObject 会记录各个 Node 的 Render 信息，比如 RenderStyle、Node 和 RenderLayer 等。当 WebKit 创建 DOM Tree 的同时也会创建 RenderObject 对象，DOM Tree 动态加入一个新节点时也会创建相应的 RenderObject 对象。

整个 Tree 的创建过程是由 NodeRenderingContext 类来完成的。WebKit 先检查 DOM 节点是否需要新的 RenderObject 对象，需要则创建或者获取一个 NodeRenderingContext 对象来创建 RenderObject 对象。NodeRenderingContext 的作用是分析设置 RenderObject 对象的父节点、兄弟节点等信息，然后插入 RenderObject 树。

RenderObject

RenderObject 需要知道 CSS 的相关属性，并且引用在 CSS 解析过程中生成的 CSSStyleSelector 提供的 RenderStyle。RenderObject 的创建过程的示意图如下：

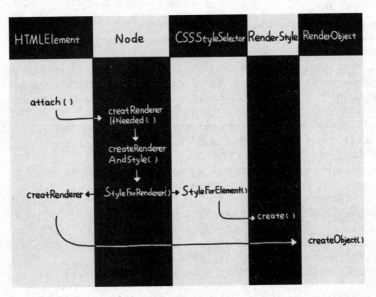

创建 RenderObject 时，Element 对象会递归调用 attach 函数，并检查是否需要创建一个新的 RenderObject。如果需要创建，则使用 NodeRenderingContext 类来创建不同 DOM 节点类型的 RenderObject。attach 函数的执行过程是在当前的 Render 节点的父节点中将当前 Render 节点插入到合适的位置，父节点再设置当前节点的兄弟节点。

RenderObject 中的主要虚函数如下：

- parent()、firstChild()、nextSibling()、previousSibling()、addChild()、removeChild()，这些是遍历和修改 Render Tree 的函数。
- layout()、style()、enclosingBox()是计算和获取布局的函数。
- isASubClass 判断是哪种子类的函数。
- paint()、repaint() 将绘制内容传入 paint()，完成绘制后，将包含坐标的绘制结果对象传入 repaint()进行渲染。

RenderObject 用来构建 Render Tree 的相关方法如下：

```
RenderElement* parent() const { return m_parent; }
bool isDescendantOf(const RenderObject*) const;

RenderObject* previousSibling() const { return m_previous; }
RenderObject* nextSibling() const { return m_next; }

// Use RenderElement versions instead.
virtual RenderObject* firstChildSlow() const { return nullptr; }
virtual RenderObject* lastChildSlow() const { return nullptr; }

RenderObject* nextInPreOrder() const;
RenderObject* nextInPreOrder(const RenderObject* stayWithin) const;
RenderObject* nextInPreOrderAfterChildren() const;
RenderObject* nextInPreOrderAfterChildren(const RenderObject* stayWithin) const;
RenderObject* previousInPreOrder() const;
RenderObject* previousInPreOrder(const RenderObject* stayWithin) const;
WEBCORE_EXPORT RenderObject* childAt(unsigned) const;
```

```
RenderObject* firstLeafChild() const;
RenderObject* lastLeafChild() const;
```

RenderObject 用来布局的相关方法如下：

```
bool needsLayout() const
{
    return m_bitfields.needsLayout() || m_bitfields
        .normalChildNeedsLayout() || m_bitfields.posChildNeedsLayout()
        || m_bitfields.needsSimplifiedNormalFlowLayout()
        || m_bitfields.needsPositionedMovementLayout();
}

bool selfNeedsLayout() const { return m_bitfields.needsLayout(); }
bool needsPositionedMovementLayout() const { return
m_bitfields.needsPositionedMovementLayout(); }
bool needsPositionedMovementLayoutOnly() const
{
    return m_bitfields.needsPositionedMovementLayout() && !m_bitfields
        .needsLayout() && !m_bitfields.normalChildNeedsLayout()
        && !m_bitfields.posChildNeedsLayout() && !m_bitfields
        .needsSimplifiedNormalFlowLayout();
}

bool posChildNeedsLayout() const { return m_bitfields.posChildNeedsLayout(); }
bool needsSimplifiedNormalFlowLayout() const { return
m_bitfields.needsSimplifiedNormalFlowLayout(); }
bool normalChildNeedsLayout() const { return
m_bitfields.normalChildNeedsLayout(); }
```

RenderBoxModelObject 是 RenderObject 的子类，是对 CSS Box 的封装，其定义了 CSS Box 的各个接口。

RenderBox 是 RenderBoxModelObject 的子类，会重载 RenderObject 和 RenderBoxModelObject 这两个对象类里的一些方法。

RenderBlock 是 RenderBox 的子类，用于封装 CSS 里 Block 的元素，RenderBlock 类里有布局 Inline、block 和定位等相关方法，有删除、插入浮动节点，还有计算浮动节点位置等方法，以及绘制相关的方法，如绘制浮动节点的方法、绘制 outline 的方法、绘制选择区的方法等。

RenderView 继承自 RenderBlock，Render Tree 的根节点是 Tree 布局和渲染的入口。

RenderInlineBox 是 RenderBoxModelObject 的子类，用于封装 CSS 里的 Inline 元素。其方法主要用于处理 Inline 的自动换多行的问题。对多行的处理，RenderInlineBox 会用 RenderInlineBoxList 类的 m_lineBoxes 持有多个 InlineBox 的行元素。

RenderStyle 可以保存 CSS 的解析结果，并且 CSS 属性会保存在 InheritedFlags 和 NonInheritedFlags 里。在 InheritedFlags 里保存了文本方向和文本对齐等属性；在 NonInheritedFlags 里保存了 float、position、overflow 等属性，以及大量的访问和设置 CSS 属性的方法。

RenderText 继承自 RenderObject 并提供文字处理的相关功能。

StyleResolver

RenderStyle 对象来自 CSS，其保存了 Render Tree 绘制需要的所有内容。StyleResolver 类负责将 CSS 转成 RenderStyle。StyleResolver 类将元素的信息保存到新建的 RenderStyle 对象中，元素信息包括标签名和类别等，它们最后被 RenderObject 类所管理和使用。

规则是怎样匹配的？这是由 ElementRuleCollector 类计算获得的，根据元素属性信息，从 DocumentRuleSets 类里获得规则集合，按照 id、class、tag 等信息获得元素的样式。

当创建 RenderObject 类时，StyleResolver 会收集样式信息给 RenderStyle 对象，这样 RenderStyle 对象就包含了完整的样式信息，代码如下：

```
PassRefPtr<RenderStyle> styleForElement(Element*, RenderStyle* parentStyle = 0,
StyleSharingBehavior = AllowStyleSharing,
    RuleMatchingBehavior = MatchAllRules, RenderRegion* regionForStyling = 0);
……

PassRefPtr<RenderStyle> pseudoStyleForElement(PseudoId, Element*, RenderStyle*
parentStyle);

PassRefPtr<RenderStyle> styleForPage(int pageIndex);
PassRefPtr<RenderStyle> defaultStyleForElement();
PassRefPtr<RenderStyle> styleForText(Text*);

static PassRefPtr<RenderStyle> styleForDocument(Document*, CSSFontSelector* =
0);
```

StyleResolver 是 Document 的子对象。当 CSS 发生变化（style 发生变化）时，就会调用 recalcStyle，recalcStyle 调用 Element::StyleForRender，Element::StyleForRender 调用 styleForElement。

styleForElement 进行 CSS 选择和匹配的具体实现代码如下：

```
initElement(element);
initForStyleResolve(element, defaultParent);
...

MatchResult matchResult;
if (matchingBehavior == MatchOnlyUserAgentRules)
    matchUARules(matchResult);
else
    matchAllRules(matchResult, matchingBehavior != MatchAllRulesExcludingSMIL);

applyMatchedProperties(matchResult, element);

void StyleResolver::matchUARules(MatchResult& result, RuleSet* rules)
{
    m_matchedRules.clear();
    result.ranges.lastUARule = result.matchedProperties.size() - 1;
    collectMatchingRules(rules, result.ranges.firstUARule, result.ranges
        .lastUARule, false);
```

```
    sortAndTransferMatchedRules(result);
}
```

sortAndTransferMatchedRules 是为了保证正确的匹配顺序。

RuleSet

matchUARules 的 RuleSet 代表 CSS 规则，代码如下：

```
p { background : red; }
```

RuleSet 的成员变量的代码如下：

```
class RuleSet ... {
...
AtomRuleMap m_idRules;
AtomRuleMap m_classRules;
AtomRuleMap m_tagRules;
AtomRuleMap m_shadowPseudoElementRules;
...
};
```

RuleSet 可以是浏览器默认的 RuleSet，或者是页面自己定义的 Author RuleSet。对于 Element 的 RuleSet 来说，只有继承 StyledElement 的内联 CSS 的 Element 才能够有 RuleSet。RuleSet 是在 StyledElement::rebuildPresentationAttributeStyle 里实现的，代码如下：

```
void StyledElement::rebuildPresentationAttributeStyle()
{
...

    RefPtr<StylePropertySet> style;
    if (cacheHash && cacheIterator->value) {
        style = cacheIterator->value->value;
        presentationAttributeCacheCleaner()
            .didHitPresentationAttributeCache();
    } else {
        style = StylePropertySet::create(isSVGElement() ? SVGAttributeMode : CSSQuirksMode);
        unsigned size = attributeCount();
        for (unsigned i = 0; i < size; ++i) {
            const Attribute* attribute = attributeItem(i);
            collectStyleForPresentationAttribute(*attribute, style.get());
        }
    }

    // ImmutableElementAttributeData doesn't store
    // presentation attribute style,
    // so make sure we have a MutableElementAttributeData.
    ElementAttributeData* attributeData = mutableAttributeData();

    attributeData->m_presentationAttributeStyleIsDirty = false;
    attributeData->setPresentationAttributeStyle(style->isEmpty() ? 0 : style);
...
```

AuthorStyle 用 StyleResolver::m_authorStyle 来保存 HTML 里所有 style tag 和 link 的

外部 CSS。当 CSS 被改变或者页面大小变化时，WebCore 会调用 Document::styleResolverChanged，接着调用 DocumentStyleCollection:: updateActiveStyleSheets，最后调用 StyleResolver::appendAuthorStyleSheets。

Render Tree 和 DOM Tree 做对应

Render Tree 和 DOM Tree 其实并不是一一对应的，比如说 head 标签或者 display 属性为 none 的，就不会出现在 Render Tree 中。

有些 RenderObject 有对应的 DOM Node，但和 DOM Tree 上的 Node 所处位置不同，比如 float 和 absolutely position 的元素，这些元素所处位置和 DOM Tree 所处位置并不相同，会根据设置的 frame 直接映射到对应的位置，如下图所示。

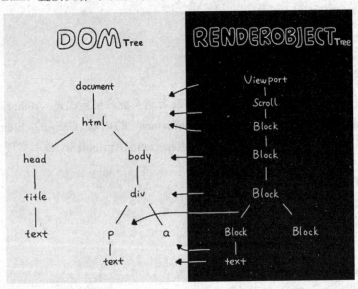

构建 Render Tree 的流程

解决样式和创建一个 Renderer 的过程称为 "attachment"。每个 Element 有一个 attach 方法，Render Tree 的创建就是从 attach 方法开始的，attachment 是同步操作的，在 Element 节点插入 DOM Tree，则会调用新 Node 的 attach 方法。在处理 html 和 body 标签时，WebCore 会创建 Render Tree 的 root。rootRender Object 对应的是 CSS 里的 containing block。top block 会包含其他所有的 block。RenderObject 是通过 RenderObject::CreateObject 静态方法创建的，创建时会计算出 Element 的 CSS Style。浏览器 window 中的显示区域 viewport 在 WebCore 里叫作 RenderView。

RenderLayer Tree

RenderLayer Tree 是基于 RenderObject Tree 建立的，RenderLayer Tree 记录多个 RenderObject 对应一个 RenderLayer 的关系。在创建完 RenderObject Tree 后，WebKit 还会创建 RenderLayer Tree，根节点是 RenderView 的实例，RenderLayer Tree 根据条件判断自己是否需要创建一个新的 RenderLayer 对象，再设置 RenderLayer 对象的父对象和兄弟对象。

RenderLayer 需要满足一定条件才会被创建，下面是创建新 RenderLayer 对象的条件：
- HTMLElement 节点对应的是 RenderBlock 节点。
- 设置了 relative、absolute、transform 的 CSS position 属性。
- 透明效果的 RenderObject 节点。
- 节点有 overflow 节点、apha mask 节点或反射效果的 RenderObject 节点。
- 设置了 CSS filter 属性的节点。
- 使用 Canvas 或 WebGL 的 RenderObject 节点。
- Video 节点对应的是 RenderObject 节点。

style 计算

构建 Render Tree 需要计算每个 RenderObject 的视觉特性。通过计算每个 Element 的 style 属性来完成任务。

Element 的 style 包含了各种来源的 style sheet，style sheet 里有 inline style Element 和 HTML 里的视觉属性，比如 bgcolor。这些属性都会转为 CSS Style 的属性。Element 的 style 来源还有浏览器默认的 style sheet，这个 style sheet 可以是用户自定义的。

我们先看一看 style 计算可能会出现的问题：
- style 的数据是个非常大的结构，因为里面有大量的 style 属性，这样的结构会有内存问题。
- 如果没有为每个 Element 查找合适的规则，就会导致性能问题。在整个规则列表里为每个 Element 查找合适的规则的消耗巨大。selector 可能会有很复杂的结构，比如 div div div div {...} 这样的代码描述形式，就会进行很多次的遍历来匹配规则。
- 应用规则涉及比较复杂的层级规则。

下面我们来说一下如何解决这些问题。

首先是共享 style 数据。WebKit 的 Node 是 style object（RenderStyle）。这些 object 可以被 Node 在某些情况下共享，这些 Node 可以匹配上规则。以下情况是可以用 style 共享的：
- Element 必须有相同的鼠标状态。
- 没有 id。
- tag name 能匹配上规则。
- class name 能匹配上规则。
- 一组映射的属性是相同的。
- 链接状态能匹配上规则。
- focus 状态能匹配上规则。
- 任何 Element 都不会被属性 selector 影响。
- Element 不能有 inline style 属性。
- 不要使用兄弟 selector。

WebKit 会把 style object 和 DOM 的 Node 相关联。通过正确的顺序将所有逻辑规则值转成计算值,比如百分比会被转换成绝对值,Rule Tree 可以使节点之间共享这些值,避免多次计算,同时节省内存。所有匹配的规则都会保存在一个 Tree 里。Tree 最低层级的 Node 有着最高的优先级。Rule Tree 是当 node style 需要时才会根据规则计算的。

可确定的样式会分配到对应的 struct 里。一些样式的信息是可以从一开始就确定的,比如 border 和 color。所有属性分为可继承属性和不可继承属性两种,可继承属性需要特别指定,默认是不可继承属性。

Rule Tree 可以缓存全部的 struct,这样的好处是如果下面的 Node 没有可用的 struct 时,则可以使用父节点缓存里的 struct。

使用 Rule Tree 计算 style context。当确定要为某个 Element 计算 style context 时,首先计算 Rule Tree 的路径或者使用一个已经缓存的路径。接着可以在新的 style context 里运用路径里的规则去填充 struct。

从路径下面的 Node 开始查找,直到找到 struct 为止。如果 struct 没有符合的 Rule Node,那么去 Rule Tree 上面找 Rule Node,找到就可以直接使用,完成优化。整个 struct 都是共享的。

如果没有找到一个 Rule Node,那么先检查 struct 是不是可继承的。如果可继承,就使用父 struct 的 context。如果不可继承,那么就使用默认的值。

下面是不同的 style rule 的来源:

- CSS Rule,可以来自 style sheet,也可以来自 style element 里的定义。

```
p {color:blue}
```

- Inline style 属性。

```
<p style="color:blue" />
```

- HTML 视觉属性。

```
<p bgcolor="blue">
```

后面两个属性属于容易匹配的 Element,可以使用 Element 作为 Map 的 key。解析 style sheet 后,通过 selector rule 会添加一些 HashMap。这些 HashMap 包括 id 的 HashMap,class name 的 HashMap,还有 tag name 的 HashMap。如果 selector 是 id,那么 rule 会被添加到 id 的 HashMap 里。

如果 selector 是 class name,rule 就会被添加到 class name 的 HashMap 里。

将 rule 添加到 HashMap 里的操作使 rule 被匹配的效率更高。完全没必要去查找每个声明,我们可以从对应的 Element 的 Map 里提取相关的 rule。

一个 style 属性可以被定义在多个 style sheet 里,那么应用 rule 的顺序就非常重要了,在 CSS 的定义里把这个顺序叫作 cascade 排序,意思是从低到高排序。

rule 设置的优先级由低到高的排列如下:

- 浏览器的设置。
- 用户的设置。
- 网页里普通的设置。

- 网页里重要的设置。
- 用户重要的设置。

计算方法如下：

- style attribute 数量 = a。
- ID attribute 数量 = b。
- 其他的 attribute 和 pseudo-classes 数量 = c。
- element name 和 pseudo-element 数量 = d。

这些 rule 匹配后就需要对优先级排序了，WebKit 使用的是冒泡排序，通过重写 > 操作符来进行排序，代码如下：

```
static bool operator >(CSSRuleData& r1, CSSRuleData& r2)
{
    int spec1 = r1.selector()->specificity();
    int spec2 = r2.selector()->specificity();
    return (spec1 == spec2) : r1.position() > r2.position() : spec1 > spec2;
}
```

3.4.10 Layout

当 Renderer 创建完并添加到 Tree 里时，Renderer 是没有位置给这些信息的，需要计算位置大小。WebKit 计算位置大小的过程称为布局计算。

HTML 使用的是流式布局模型，即一般情况下一次就可以按照从左到右或从上到下算出布局需要的值。在特殊情况下，需要 HTML table 的次数可能不止一次。

Layout 为递归过程，以 FrameView::layout 为起始，调用 Render Tree 根节点的 Layout，对应 HTML 里的 Element，Layout 持续递归 frame hierarchy 计算每个 Renderer 的几何信息。root 的 position 是 0,0，把浏览器里可见部分的尺寸叫作 viewport。每个 Renderer 都有 Layout 或 Reflow 方法，每个 Renderer 都会调用子视图的 Layout 方法。FrameView::layout 可以被布局的 WebViewCore::layout 触发。

Layout 处理过程

完成 Layout 递归计算的过程之后，即可开始布局，执行 FrameView::layout 了。FrameView::layout 被执行之前的方法堆栈如下所示：

```
FrameView:: layout( bool allowSubtree)
Document:: implicitClose()
FrameLoader:: checkCallImplicitClose()
FrameLoader:: checkCompleted()
FrameLoader:: finishedParsing()
Document:: finishedParsing()
HTMLParser:: finished()
HTMLTokenizer:: end()
HTMLTokenizer:: finish()
Document:: finishParsing()
FrameLoader:: endIfNotLoadingMainResource()
FrameLoader:: end()
```

```
DocumentLoader::finishedLoading()
FrameLoader::finishedLoading()
MainResourceLoader::didFinishLoading()
ResourceLoader::didFinishLoading( WebCore::ResourceHandle * h)
```

Layout 会先暂停 Layout 定时器的实现，然后更新样式，获取到 RenderView，然后调用它的 Layout 方法开始布局，相关的代码如下：

```
void FrameView::layout(bool allowSubtree)
{
    ASSERT_WITH_SECURITY_IMPLICATION(!frame().document()
      ->inRenderTreeUpdate());

    LOG(Layout, "FrameView %p (%dx%d) layout, main frameview %d, allowSubtree=%d",
this, size().width(), size().height(), frame().isMainFrame(), allowSubtree);
    if (isInRenderTreeLayout()) {
        LOG(Layout, " in layout, bailing");
        return;
    }

    if (layoutDisallowed()) {
        LOG(Layout, " layout is disallowed, bailing");
        return;
    }

    // Protect the view from being deleted during layout (in recalcStyle).
    Ref<FrameView> protectedThis(*this);

    // Many of the tasks performed during layout can cause
    // this function to be re-entered,
    // so save the layout phase now and restore it on exit.
    SetForScope<LayoutPhase> layoutPhaseRestorer(m_layoutPhase, InPreLayout);

    // Every scroll that happens during layout is programmatic.
    SetForScope<bool> changeInProgrammaticScroll(m_inProgrammaticScroll, true);

    bool inChildFrameLayoutWithFrameFlattening =
isInChildFrameWithFrameFlattening();

    if (inChildFrameLayoutWithFrameFlattening) {
        if (!m_frameFlatteningViewSizeForMediaQuery) {
            LOG_WITH_STREAM(MediaQueries, stream << "FrameView " << this << "
snapshotting size " << ScrollView::layoutSize() << " for media queries");
            m_frameFlatteningViewSizeForMediaQuery = ScrollView
                                                       ::layoutSize();
        }
        startLayoutAtMainFrameViewIfNeeded(allowSubtree);
        RenderElement* root = m_layoutRoot ? m_layoutRoot : frame()
                         .document()->renderView();
        if (!root || !root->needsLayout())
            return;
    }

    TraceScope tracingScope(LayoutStart, LayoutEnd);
```

```cpp
#if PLATFORM(IOS)
    if (updateFixedPositionLayoutRect())
        allowSubtree = false;
#endif
    m_layoutTimer.stop();
    m_delayedLayout = false;
    m_setNeedsLayoutWasDeferred = false;

    ASSERT(!isPainting());
    if (isPainting())
        return;

    InspectorInstrumentationCookie cookie = InspectorInstrumentation
                                            ::willLayout(frame());
    AnimationUpdateBlock animationUpdateBlock(&frame().animation());

    if (!allowSubtree && m_layoutRoot)
        convertSubtreeLayoutToFullLayout();

    ASSERT(frame().view() == this);
    ASSERT(frame().document());

    Document& document = *frame().document();
    ASSERT(document.pageCacheState() == Document::NotInPageCache);
    {
        SetForScope<bool> changeSchedulingEnabled(m_layoutSchedulingEnabled, false);

        if (!m_nestedLayoutCount && !m_inSynchronousPostLayout && m_postLayoutTasksTimer.isActive() && !inChildFrameLayoutWithFrameFlattening) {
            // This is a new top-level layout. If there are any remaining tasks from the previous
            // layout, finish them now.
            SetForScope<bool> inSynchronousPostLayoutChange(m_inSynchronousPostLayout, true);
            performPostLayoutTasks();
        }

        m_layoutPhase = InPreLayoutStyleUpdate;

        // Viewport-dependent media queries
        // may cause us to need completely different style information.
        auto* styleResolver = document.styleScope().resolverIfExists();
        if (!styleResolver || styleResolver->hasMediaQueriesAffectedByViewportChange()) {
            LOG(Layout, " hasMediaQueriesAffectedByViewportChange, enqueueing style recalc");
            document.styleScope().didChangeStyleSheetEnvironment();
            // FIXME: This instrumentation event is not strictly accurate since cached media query results do not persist across StyleResolver rebuilds.
            InspectorInstrumentation::mediaQueryResultChanged(document);
        }
        document.evaluateMediaQueryList();
```

```cpp
    // If there is any pagination to apply,
    // it will affect the RenderView's style, so we should
    // take care of that now.
    applyPaginationToViewport();
    // Always ensure our style info is up-to-date.
    // This can happen in situations where
    // the layout beats any sort of style recalc update that needs to occur.
    document.updateStyleIfNeeded();
    // If there is only one ref to this view left,
    // then its going to be destroyed as soon as we exit,
    // so there's no point to continuing to layout
    if (hasOneRef())
        return;

    // Close block here so we can set up the font cache purge preventer,
    // which we will still
    // want in scope even after we want m_layoutSchedulingEnabled
    // to be restored again.
    // The next block sets m_layoutSchedulingEnabled back
    // to false once again.
}

m_layoutPhase = InPreLayout;

RenderLayer* layer = nullptr;
bool subtree = false;
RenderElement* root = nullptr;

++m_nestedLayoutCount;

{
    SetForScope<bool> changeSchedulingEnabled(
      m_layoutSchedulingEnabled, false);

    autoSizeIfEnabled();
    root = m_layoutRoot ? m_layoutRoot : document.renderView();
    if (!root)
        return;
    subtree = m_layoutRoot;

    if (!m_layoutRoot) {
        auto* body = document.bodyOrFrameset();
        if (body && body->renderer()) {
            if (is<HTMLFrameSetElement>(*body)
                && !frameFlatteningEnabled()) {
                body->renderer()->setChildNeedsLayout();
            } else if (is<HTMLBodyElement>(*body)) {
                if (!m_firstLayout && m_size.height() != layoutHeight()
                    && body->renderer()->enclosingBox()
                    .stretchesToViewport())
                    body->renderer()->setChildNeedsLayout();
            }
        }
    }
```

```cpp
    #if !LOG_DISABLED
            if (m_firstLayout && !frame().ownerElement())
                LOG(Layout, "FrameView %p elapsed time before first layout: %.3fs\n",
this, document.timeSinceDocumentCreation().value());
    #endif
        }

        m_needsFullRepaint = !subtree && (m_firstLayout ||
downcast<RenderView>(*root).printing());

        if (!subtree) {
            ScrollbarMode hMode;
            ScrollbarMode vMode;
            calculateScrollbarModesForLayout(hMode, vMode);

            if (m_firstLayout || (hMode != horizontalScrollbarMode() || vMode !=
verticalScrollbarMode())) {
                if (m_firstLayout) {
                    setScrollbarsSuppressed(true);

                    m_firstLayout = false;
                    m_firstLayoutCallbackPending = true;
                    m_lastViewportSize = sizeForResizeEvent();
                    m_lastZoomFactor = root->style().zoom();

                    // Set the initial vMode to AlwaysOn if we're auto.
                    if (vMode == ScrollbarAuto)
                        // This causes a vertical scrollbar to appear.
                        setVerticalScrollbarMode(ScrollbarAlwaysOn);
                    // Set the initial hMode to AlwaysOff if we're auto.
                    if (hMode == ScrollbarAuto)
                        // This causes a horizontal scrollbar to disappear.
                        setHorizontalScrollbarMode(ScrollbarAlwaysOff);
                    Page* page = frame().page();
                    if (page && page->expectsWheelEventTriggers())
                        scrollAnimator().setWheelEventTestTrigger(page
                           ->testTrigger());
                    setScrollbarModes(hMode, vMode);
                    setScrollbarsSuppressed(false, true);
                } else
                    setScrollbarModes(hMode, vMode);
            }

            LayoutSize oldSize = m_size;
            m_size = layoutSize();

            if (oldSize != m_size) {
                LOG(Layout, " layout size changed from %.3fx%.3f to %.3fx%.3f",
oldSize.width().toFloat(), oldSize.height().toFloat(), m_size.width().toFloat(),
m_size.height().toFloat());
                m_needsFullRepaint = true;
                if (!m_firstLayout) {
                    RenderBox* rootRenderer = document.documentElement() ?
document.documentElement()->renderBox() : nullptr;
```

```cpp
                auto* body = document.bodyOrFrameset();
                RenderBox* bodyRenderer = rootRenderer && body ? 
body->renderBox() : nullptr;
                if (bodyRenderer && bodyRenderer->stretchesToViewport())
                    bodyRenderer->setChildNeedsLayout();
                else if (rootRenderer && rootRenderer
                  ->stretchesToViewport())
                    rootRenderer->setChildNeedsLayout();
            }
        }

        m_layoutPhase = InPreLayout;
    }

    layer = root->enclosingLayer();
    SubtreeLayoutStateMaintainer subtreeLayoutStateMaintainer(
      m_layoutRoot);

    RenderView::RepaintRegionAccumulator repaintRegionAccumulator(
      &root->view());

    ASSERT(m_layoutPhase == InPreLayout);
    m_layoutPhase = InRenderTreeLayout;

    forceLayoutParentViewIfNeeded();

    ASSERT(m_layoutPhase == InRenderTreeLayout);
#ifndef NDEBUG
    RenderTreeNeedsLayoutChecker checker(*root);
#endif
    root->layout();
    ASSERT(!root->view().renderTreeIsBeingMutatedInternally());

#if ENABLE(TEXT_AUTOSIZING)
    if (frame().settings().textAutosizingEnabled() && !root
      ->view().printing()) {
        float minimumZoomFontSize = frame().settings()
          .minimumZoomFontSize();
        float textAutosizingWidth = frame().page() ? frame().page()
          ->textAutosizingWidth() : 0;
        if (int overrideWidth = frame().settings()
          .textAutosizingWindowSizeOverride().width())
            textAutosizingWidth = overrideWidth;

        LOG(TextAutosizing, "Text Autosizing: minimumZoomFontSize=%.2f 
textAutosizingWidth=%.2f", minimumZoomFontSize, textAutosizingWidth);

        if (minimumZoomFontSize && textAutosizingWidth) {
            root->adjustComputedFontSizesOnBlocks(minimumZoomFontSize, 
textAutosizingWidth);
            if (root->needsLayout())
                root->layout();
        }
    }
```

```
    #endif

        ASSERT(m_layoutPhase == InRenderTreeLayout);
        m_layoutRoot = nullptr;
        // Close block here to end the scope of changeSchedulingEnabled
        // and SubtreeLayoutStateMaintainer
    }

    m_layoutPhase = InViewSizeAdjust;

    bool neededFullRepaint = m_needsFullRepaint;

    if (!subtree && !downcast<RenderView>(*root).printing()) {
        adjustViewSize();
        // FIXME: Firing media query callbacks synchronously on
        // nested frames could produced a detached FrameView here by
        // navigating away from the current document
        // (see webkit.org/b/173329)
        if (hasOneRef())
            return;
    }

    m_layoutPhase = InPostLayout;

    m_needsFullRepaint = neededFullRepaint;

    // Now update the positions of all layers
    if (m_needsFullRepaint)
        root->view().repaintRootContents();

    root->view().releaseProtectedRenderWidgets();

    ASSERT(!root->needsLayout());

    layer->updateLayerPositionsAfterLayout(renderView()->layer(),
updateLayerPositionFlags(layer, subtree, m_needsFullRepaint));

    updateCompositingLayersAfterLayout();

    m_layoutPhase = InPostLayerPositionsUpdatedAfterLayout;

    m_layoutCount++;

#if PLATFORM(COCOA) || PLATFORM(WIN) || PLATFORM(GTK)
    if (AXObjectCache* cache = root->document().existingAXObjectCache())
        cache->postNotification(root, AXObjectCache::AXLayoutComplete);
#endif

#if ENABLE(DASHBOARD_SUPPORT)
    updateAnnotatedRegions();
#endif

#if ENABLE(IOS_TOUCH_EVENTS)
    document.setTouchEventRegionsNeedUpdate();
```

```
#endif

    updateCanBlitOnScrollRecursively();

    handleDeferredScrollUpdateAfterContentSizeChange();

    handleDeferredScrollbarsUpdateAfterDirectionChange();

    if (document.hasListenerType(Document::OVERFLOWCHANGED_LISTENER))
        updateOverflowStatus(layoutWidth() < contentsWidth(),
        layoutHeight() < contentsHeight());

    frame().document()->markers().invalidateRectsForAllMarkers();

    if (!m_postLayoutTasksTimer.isActive()) {
        if (!m_inSynchronousPostLayout) {
            if (inChildFrameLayoutWithFrameFlattening)
                updateWidgetPositions();
            else {
                SetForScope<bool> inSynchronousPostLayoutChange(
                    m_inSynchronousPostLayout, true);
                performPostLayoutTasks(); // Calls resumeScheduledEvents().
            }
        }

        if (!m_postLayoutTasksTimer.isActive() && (needsLayout() || m_inSynchronousPostLayout || inChildFrameLayoutWithFrameFlattening)) {
            // If we need layout or are already in a synchronous
            // call to postLayoutTasks(),
            // defer widget updates and event dispatch until after we return.
            // postLayoutTasks() can make us need to update again,
            // and we can get stuck in a nasty cycle unless
            // we call it through the timer here.
            m_postLayoutTasksTimer.startOneShot(0_s);
        }
        if (needsLayout())
            layout();
    }

    InspectorInstrumentation::didLayout(cookie, *root);
    DebugPageOverlays::didLayout(frame());

    --m_nestedLayoutCount;
}
```

在 RenderView 的 Layout 里调用 layoutContent 方法。这个方法的实现如下:

```
void RenderView::layoutContent(const LayoutState& state)
{
    UNUSED_PARAM(state);
    ASSERT(needsLayout());

    RenderBlockFlow::layout();
    if (hasRenderNamedFlowThreads())
```

```
        flowThreadController().layoutRenderNamedFlowThreads();
#ifndef NDEBUG
    checkLayoutState(state);
#endif
}
```

这个方法的实现很简单，主要通过 RenderBlockFlow 进行布局。RenderBlockFlow 是 RenderBlock 的子类，布局的方法是在 RenderBlock 里实现的，其直接调用了 layoutBlock 方法。这个方法是在 RenderBlockFlow 里重载实现的。下面为 RenderBlockFlow 的实现：

```
void RenderBlockFlow::layoutBlock(bool relayoutChildren, LayoutUnit pageLogicalHeight)
{
    ASSERT(needsLayout());
    if (!relayoutChildren && simplifiedLayout())
        return;

    LayoutRepainter repainter(*this, checkForRepaintDuringLayout());
    if (recomputeLogicalWidthAndColumnWidth())
        relayoutChildren = true;

    rebuildFloatingObjectSetFromIntrudingFloats();

    LayoutUnit previousHeight = logicalHeight();
    // FIXME: should this start out
    // as borderAndPaddingLogicalHeight() + scrollbarLogicalHeight()
    // for consistency with other render classes
    setLogicalHeight(0);

    bool pageLogicalHeightChanged = false;
    checkForPaginationLogicalHeightChange(relayoutChildren, pageLogicalHeight, pageLogicalHeightChanged);

    const RenderStyle& styleToUse = style();
    LayoutStateMaintainer statePusher(view(), *this, locationOffset(), hasTransform() || hasReflection() || styleToUse.isFlippedBlocksWritingMode(), pageLogicalHeight, pageLogicalHeightChanged);

    preparePaginationBeforeBlockLayout(relayoutChildren);
    if (!relayoutChildren)
        relayoutChildren = namedFlowFragmentNeedsUpdate();

    // We use four values, maxTopPos, maxTopNeg, maxBottomPos
    // and maxBottomNeg, to track our current maximal positive
    // and negative margins. These values are used when we are collapsed
    // with adjacent blocks, so for example, if you have block A and B
    // collapsing together, then you'd take the maximal positive margin
    // from both A and B and subtract it from the maximal negative margin
    // from both A and B to get the true collapsed margin
    // This algorithm is recursive, so when we finish layout()
    // our block knows its current maximal positive/negative values
    // Start out by setting our margin values to our current margins
    // Table cells have no margins
```

```cpp
        // so we don't fill in the values for table cells
        bool isCell = isTableCell();
        if (!isCell) {
            initMaxMarginValues();

            setHasMarginBeforeQuirk(styleToUse.hasMarginBeforeQuirk());
            setHasMarginAfterQuirk(styleToUse.hasMarginAfterQuirk());
            setPaginationStrut(0);
        }

        LayoutUnit repaintLogicalTop = 0;
        LayoutUnit repaintLogicalBottom = 0;
        LayoutUnit maxFloatLogicalBottom = 0;
        if (!firstChild() && !isAnonymousBlock())
            setChildrenInline(true);
        if (childrenInline())
            layoutInlineChildren(relayoutChildren, repaintLogicalTop, repaintLogicalBottom);
        else
            layoutBlockChildren(relayoutChildren, maxFloatLogicalBottom);

        // Expand our intrinsic height to encompass floats
        LayoutUnit toAdd = borderAndPaddingAfter() + scrollbarLogicalHeight();
        if (lowestFloatLogicalBottom() > (logicalHeight() - toAdd) && createsNewFormattingContext())
            setLogicalHeight(lowestFloatLogicalBottom() + toAdd);

        if (relayoutForPagination(statePusher) || relayoutToAvoidWidows(statePusher)) {
            ASSERT(!shouldBreakAtLineToAvoidWidow());
            return;
        }

        // Calculate our new height  LayoutUnit oldHeight = logicalHeight();
        LayoutUnit oldClientAfterEdge = clientLogicalBottom();

        // Before updating the final size of the flow thread
        // make sure a forced break is applied after the content
        // This ensures the size information is correctly computed for
        // the last auto-height region receiving content
        if (is<RenderFlowThread>(*this))
            downcast<RenderFlowThread>(*this)
              .applyBreakAfterContent(oldClientAfterEdge);

        updateLogicalHeight();
        LayoutUnit newHeight = logicalHeight();
        if (oldHeight != newHeight) {
            if (oldHeight > newHeight && maxFloatLogicalBottom > newHeight && !childrenInline()) {
                // One of our children's floats may have
                // become an overhanging float for us. We need to look for it
                for (auto& blockFlow : childrenOfType<RenderBlockFlow>(*this)) {
                    if (blockFlow.isFloatingOrOutOfFlowPositioned())
                        continue;
```

```cpp
                if (blockFlow.lowestFloatLogicalBottom() + blockFlow
                    .logicalTop() > newHeight)
                    addOverhangingFloats(blockFlow, false);
            }
        }
    }

    bool heightChanged = (previousHeight != newHeight);
    if (heightChanged)
        relayoutChildren = true;

    layoutPositionedObjects(relayoutChildren || isDocumentElementRenderer());

    //Add overflow from children (unless we're multi-column
    //since in that case all our child overflow is clipped anyway)
    computeOverflow(oldClientAfterEdge);

    statePusher.pop();

    fitBorderToLinesIfNeeded();

    if (view().layoutState()->m_pageLogicalHeight)
        setPageLogicalOffset(view().layoutState()->pageLogicalOffset(
            this, logicalTop()));

    updateLayerTransform();

    // Update our scroll information if we're
    // overflow:auto/scroll/hidden now that we know if we overflow or not.
    updateScrollInfoAfterLayout();

    // FIXME: This repaint logic should be moved into
    // a separate helper function
    // Repaint with our new bounds if they are different from our old bounds
    bool didFullRepaint = repainter.repaintAfterLayout();
    if (!didFullRepaint && repaintLogicalTop != repaintLogicalBottom &&
(styleToUse.visibility() == VISIBLE || enclosingLayer()->hasVisibleContent())) {
        // FIXME: We could tighten up the left and right
        // invalidation points if we let layoutInlineChildren fill them
        // in based off the particular lines it had to lay out
        // We wouldn't need the hasOverflowClip() hack in that case either
        LayoutUnit repaintLogicalLeft = logicalLeftVisualOverflow();
        LayoutUnit repaintLogicalRight = logicalRightVisualOverflow();
        if (hasOverflowClip()) {
            // If we have clipped overflow, we should use layout overflow
            // as well, since visual overflow from lines didn't propagate
            // to our block's overflow
            // Note the old code did this as well but even for overflow:visible
            // The addition of hasOverflowClip()
            // at least tightens up the hack a bit
            // layoutInlineChildren should be patched to compute
            // the entire repaint rect
            repaintLogicalLeft = std::min(repaintLogicalLeft,
logicalLeftLayoutOverflow());
```

```
                repaintLogicalRight = std::max(repaintLogicalRight,
logicalRightLayoutOverflow());
        }

        LayoutRect repaintRect;
        if (isHorizontalWritingMode())
            repaintRect = LayoutRect(repaintLogicalLeft, repaintLogicalTop,
repaintLogicalRight - repaintLogicalLeft, repaintLogicalBottom - repaintLogicalTop);
        else
            repaintRect = LayoutRect(repaintLogicalTop, repaintLogicalLeft,
repaintLogicalBottom - repaintLogicalTop, repaintLogicalRight - repaintLogicalLeft);

        if (hasOverflowClip()) {
            // Adjust repaint rect for scroll offset
            repaintRect.moveBy(-scrollPosition());

            // Don't allow this rect to spill out of our overflow box
            repaintRect.intersect(LayoutRect(LayoutPoint(), size()));
        }

        // Make sure the rect is still non-empty
        // after intersecting for overflow above
        if (!repaintRect.isEmpty()) {
            // We need to do a partial repaint of our content.
            repaintRectangle(repaintRect);
            if (hasReflection())
                repaintRectangle(reflectedRect(repaintRect));
        }
    }

    clearNeedsLayout();
}
```

以上方法里有两个分支，一个是 layoutInlineChildren，另一个是 layoutBlockChildren。由于 Inline 属于动态调整高度，所以 layoutBlockChildren 比 Block 的实现要复杂得多，layoutInlineChildren 是 Inline 布局的入口。

Layout Tree 的创建

Parsing 完成后会触发 Layout Tree。Layout Tree 的作用是存储需要绘制的数据，Layout 的作用是处理 Node 在页面上的大小和位置。Chrome、Firefox 和 Android 使用的是 Skia 开源的 2D 图形库做的底层 Paint 引擎。代码如下：

```
void Document::finishedParsing() {
    updateStyleAndLayoutTree();
}
```

每个 Node 都会创建一个 LayoutObject，代码如下：

```
LayoutObject* newLayoutObject = m_node->createLayoutObject(style);
parentLayoutObject->addChild(newLayoutObject, nextLayoutObject);
```

Layout 值的计算

在 Layout Tree 创建完成后就开始计算 Layout 的值，比如 width、margin 等。 Block 类型节点的位置计算是在 RenderBlockFlow 的 layoutBlockChild 方法里实现的，该方法主要是先确定节点的 top 坐标，然后计算布局变化后的 margin、float 和 paging 等。根据这些计算出的值，再确定出新的 top 坐标。最后使用 determineLogicalLeftPositionForChild 得到 left 坐标。具体实现如下：

```
void RenderBlockFlow::layoutBlockChild(RenderBox& child, MarginInfo& marginInfo,
LayoutUnit& previousFloatLogicalBottom, LayoutUnit& maxFloatLogicalBottom)
{
    LayoutUnit oldPosMarginBefore = maxPositiveMarginBefore();
    LayoutUnit oldNegMarginBefore = maxNegativeMarginBefore();

    // The child is a normal flow object. Compute the margins we will use for
collapsing now
    child.computeAndSetBlockDirectionMargins(*this);

    // Try to guess our correct logical top position
    // In most cases this guess will
    // be correct. Only if we're wrong (when we compute the real logical top position)
    // will we have to potentially relayout
    LayoutUnit estimateWithoutPagination;
    LayoutUnit logicalTopEstimate = estimateLogicalTopPosition(child, marginInfo,
estimateWithoutPagination);

    // Cache our old rect so that we can dirty the proper repaint rects
    // if the child moves
    LayoutRect oldRect = child.frameRect();
    LayoutUnit oldLogicalTop = logicalTopForChild(child);

#if !ASSERT_DISABLED
    LayoutSize oldLayoutDelta = view().layoutDelta();
#endif
    // Position the child as though it didn't collapse with the top
    setLogicalTopForChild(child, logicalTopEstimate, ApplyLayoutDelta);
    estimateRegionRangeForBoxChild(child);

    RenderBlockFlow* childBlockFlow = is<RenderBlockFlow>(child) ?
&downcast<RenderBlockFlow>(child) : nullptr;
    bool markDescendantsWithFloats = false;
    if (logicalTopEstimate != oldLogicalTop && !child.avoidsFloats() &&
childBlockFlow && childBlockFlow->containsFloats())
        markDescendantsWithFloats = true;
    else if (UNLIKELY(logicalTopEstimate.mightBeSaturated()))
        // logicalTopEstimate, returned by estimateLogicalTopPosition
        // might be saturated for very large elements
        // If it does the comparison with oldLogicalTop might yield a
        // false negative as adding and removing margins
        // borders etc from a saturated number might yield incorrect results
        // If this is the case always mark for layout
        markDescendantsWithFloats = true;
```

```
        else if (!child.avoidsFloats() || child.shrinkToAvoidFloats()) {
            // If an element might be affected by the presence of floats
            // then always mark it for layout
            LayoutUnit fb = std::max(previousFloatLogicalBottom,
lowestFloatLogicalBottom());
            if (fb > logicalTopEstimate)
                markDescendantsWithFloats = true;
        }

        if (childBlockFlow) {
            if (markDescendantsWithFloats)
                childBlockFlow->markAllDescendantsWithFloatsForLayout();
            if (!child.isWritingModeRoot())
                previousFloatLogicalBottom = std::max(
                    previousFloatLogicalBottom, oldLogicalTop + childBlockFlow
                    ->lowestFloatLogicalBottom());
        }

        child.markForPaginationRelayoutIfNeeded();

        bool childHadLayout = child.everHadLayout();
        bool childNeededLayout = child.needsLayout();
        if (childNeededLayout)
            child.layout();

        // Cache if we are at the top of the block right now
        bool atBeforeSideOfBlock = marginInfo.atBeforeSideOfBlock();

        // Now determine the correct ypos based off examination of
        // collapsing margin values
        LayoutUnit logicalTopBeforeClear = collapseMargins(child, marginInfo);

        // Now check for clear
        LayoutUnit logicalTopAfterClear = clearFloatsIfNeeded(child, marginInfo,
oldPosMarginBefore, oldNegMarginBefore, logicalTopBeforeClear);

        bool paginated = view().layoutState()->isPaginated();
        if (paginated)
            logicalTopAfterClear = adjustBlockChildForPagination(
                logicalTopAfterClear, estimateWithoutPagination, child,
                atBeforeSideOfBlock && logicalTopBeforeClear
                == logicalTopAfterClear);

        setLogicalTopForChild(child, logicalTopAfterClear, ApplyLayoutDelta);

        // Now we have a final top position
        // See if it really does end up being different from our estimate
        // clearFloatsIfNeeded can also mark the child as needing a layout
        // even though we didn't move. This happens when collapseMargins
        // dynamically adds overhanging floats
        // because of a child with negative margins
        if (logicalTopAfterClear != logicalTopEstimate || child.needsLayout()
            || (paginated && childBlockFlow && childBlockFlow
            ->shouldBreakAtLineToAvoidWidow())) {
```

```
        if (child.shrinkToAvoidFloats()) {
            // The child's width depends on the line width
            // When the child shifts to clear an item, its width can
            // change (because it has more available line width)
            // So mark the item as dirty
            child.setChildNeedsLayout(MarkOnlyThis);
        }

        if (childBlockFlow) {
            if (!child.avoidsFloats() && childBlockFlow->containsFloats())
                childBlockFlow->markAllDescendantsWithFloatsForLayout();
            child.markForPaginationRelayoutIfNeeded();
        }
    }

    if (updateRegionRangeForBoxChild(child))
        child.setNeedsLayout(MarkOnlyThis);

    // In case our guess was wrong, relayout the child
    child.layoutIfNeeded();

    // We are no longer at the top of the block if we encounter
    // a non-empty child
    // This has to be done after checking for clear
    // so that margins can be reset if a clear occurred
    if (marginInfo.atBeforeSideOfBlock() && !child
      .isSelfCollapsingBlock())
        marginInfo.setAtBeforeSideOfBlock(false);

    // Now place the child in the correct left position
    determineLogicalLeftPositionForChild(child, ApplyLayoutDelta);

    // Update our height now that the child has been placed in the correct position
    setLogicalHeight(logicalHeight() +
logicalHeightForChildForFragmentation(child));
    if (mustSeparateMarginAfterForChild(child)) {
        setLogicalHeight(logicalHeight() + marginAfterForChild(child));
        marginInfo.clearMargin();
    }
    // If the child has overhanging floats that intrude into following
    // siblings (or possibly out of this block), then the parent gets
    // notified of the floats now
    if (childBlockFlow && childBlockFlow->containsFloats())
        maxFloatLogicalBottom = std::max(maxFloatLogicalBottom,
addOverhangingFloats(*childBlockFlow, !childNeededLayout));

    LayoutSize childOffset = child.location() - oldRect.location();
    if (childOffset.width() || childOffset.height()) {
        view().addLayoutDelta(childOffset);

        // If the child moved, we have to repaint it as well as any
        // floating/positioned descendants. An exception is if
        // we need a layout. In this case, we know we're going to
        // repaint ourselves (and the child) anyway
```

```
        if (childHadLayout && !selfNeedsLayout() && child
          .checkForRepaintDuringLayout())
            child.repaintDuringLayoutIfMoved(oldRect);
    }

    if (!childHadLayout && child.checkForRepaintDuringLayout()) {
        child.repaint();
        child.repaintOverhangingFloats(true);
    }

    if (paginated) {
        if (RenderFlowThread* flowThread = flowThreadContainingBlock())
            flowThread->flowThreadDescendantBoxLaidOut(&child);
        // Check for an after page/column break
        LayoutUnit newHeight = applyAfterBreak(child, logicalHeight(),
marginInfo);
        if (newHeight != height())
            setLogicalHeight(newHeight);
    }

    ASSERT(view().layoutDeltaMatches(oldLayoutDelta));
}
```

判断值的类型是固定值还是百分比，代码如下：

```
switch (length.type()) {
  case Fixed:
    return LayoutUnit(length.value());
  case Percent:
    return LayoutUnit(
        static_cast<float>(maximumValue * length.percent() / 100.0f));
}
```

计算 margin 的值，代码如下：

```
// CSS 2.1: "If both 'margin-left' and 'margin-right' are 'auto',
// their used values are equal. This horizontally centers the element
// with respect to the edges of the containing block."
const ComputedStyle& containingBlockStyle = containingBlock->styleRef();
if (marginStartLength.isAuto() && marginEndLength.isAuto()) {
  LayoutUnit centeredMarginBoxStart = std::max(
      LayoutUnit(),
      (availableWidth - childWidth) / 2);
  marginStart = centeredMarginBoxStart;
  marginEnd = availableWidth - childWidth - marginStart;
  return;
}
```

Box Model 里的设置如下：

```
m_frameRect.setWidth(width);
m_marginBox.setStart(marginLeft);
```

经过计算后的 Render Tree 具有布局信息，代码如下：

```
<html>
```

```
<body>
<p>First line.<br>Second one.</p>
</body>
</html>
```

计算 Layout 后带布局信息的 Render Tree,代码如下:

```
RenderBlock {HTML} at (0, 0) size 640x480
|— RenderBody {BODY} at (0, 80) size 640x480 [bgcolor=# FFFFFF]
| |— RenderBlock {P} at (0, 0) size 640x80
| | |— RenderText {#text} at (0, 0) size 48x24 "First line."
| | |— RenderBR {BR} at (20, 20) size 0x0
| | |— RenderText {#text} at (0, 24) size 48x24 "Second one."
```

Box Model 的数据结构

所有的 Element 都有 display 属性,用来确定显示类型。属性值有:

- block:生成一个 block box。
- inline:生成一个或多个 inline box。
- none:不生成 box。

display 属性默认是 inline,但是浏览器会设置其他默认值,比如 div 的默认设置是 block。
定位有三种方案:

- normal:由 Render Tree 决定位置。
- float:首先按照正常流程来排列,然后尽可能地往左或往右移动。
- absolute:让 Element 在 Render Tree 中的位置和在 DOM Tree 中的位置不一样。

定位方案是根据 position 属性和 float 属性来决定的,static 属性和 relative 属性表示排列方式正常, absolute 属性和 fixed 属性表示按 absolute 来确定位置。

对于 float 的处理,首先要判断宽度是否需要适配内容,代码如下:

```
bool LayoutBox::sizesLogicalWidthToFitContent(
    const Length& logicalWidth) const {
  if (isFloating() || isInlineBlockOrInlineTable())
    return true;
  ...
}
```

如果是 float 或者 inline-block ,那么宽度是由内容多少来决定的,行内的内容会将间隔控制成一个空格,前后空格和多余空格都会被忽略。

关于 float:left 的计算,代码如下:

```
while (logicalRightOffsetForPositioningFloat(
        logicalTopOffset, logicalRightOffset, &heightRemainingRight) -
        floatLogicalLeft <
    floatLogicalWidth) {
  logicalTopOffset +=
    std::min<LayoutUnit>(heightRemainingLeft, heightRemainingRight);
  floatLogicalLeft = logicalLeftOffsetForPositioningFloat(
    logicalTopOffset, logicalLeftOffset, &heightRemainingLeft);
}
```

```
}
floatLogicalLeft = std::max(
    logicalLeftOffset - borderAndPaddingLogicalLeft(), floatLogicalLeft);
```

block box 在浏览器的 window 里有自己的矩形。inline box 会在一个 block 里，inline box 里面不会有 block。block 是按照垂直顺序排列的，inline 是按照水平顺序排列的。inline box 可以被包含在另一个 line box 里，line 的高度至少和最高的 box 的高度一样，甚至更高，当 box 用的是 baseline 时，这意味着多个元素底部对齐。如果容器宽度不够，那么 inline 会变成多行。

position 有多种类型，如下：

- relative：相对定位，先按照正常的样式计算得到位置信息，然后根据设置的值再移动。
- float：Element 会被移动到 block 的左边或者 block 的右边。

absolute 和 fixed：不会参与 normal flow，只按照属性值进行精确定位，absolute 是相对于容器进行定位的，fixed 是相对于 view port 进行定位的。fixed box 不会随着 document 的滚动而移动的。

通过指定 CSS 的 z-index 属性能够达到分层的效果。z-index 表示了 box 的第三个维度，位置按照 z 轴来定。这些 box 会被分成 stack，即 stacking context，每个栈顶的 box 更容易让用户看到。stack 根据 z-index 的属性排序。

在 border 内的区域及 border 本身使用 m_frameRect 对象表示。不同方法通过不同的计算能够获取到不同的值，比如 clientWidth 的计算方法，代码如下：

```
// More IE extensions. clientWidth and clientHeight represent the interior
// of an object excluding border and scrollbar.
LayoutUnit LayoutBox::clientWidth() const {
  return m_frameRect.width() - borderLeft() - borderRight() -
      verticalScrollbarWidth();
}
```

offsetWidth 是 frameRect 的宽度，代码如下：

```
// IE extensions. Used to calculate offsetWidth/Height.
LayoutUnit offsetWidth() const override { return m_frameRect.width(); }
LayoutUnit offsetHeight() const override { return m_frameRect.height(); }
```

margin 区域是用 LayoutRectOutsets 表示的，代码如下：

```
LayoutUnit m_top;
LayoutUnit m_right;
LayoutUnit m_bottom;
LayoutUnit m_left;
```

位置的信息，即 x 和 y 是通过下面两个函数计算得到的，代码如下：

```
// 根据 margin 得到 y 值
LayoutUnit logicalTopBeforeClear =
    collapseMargins(child, layoutInfo, childIsSelfCollapsing,
                childDiscardMarginBefore, childDiscardMarginAfter);
```

```
// 根据 margin 得到 x 值
determineLogicalLeftPositionForChild(child);
```

对 x 和 y 进行计算是一个递归过程，由子元素到父元素。这样做的原因是有的父元素的高是由子元素支撑的，所以需要先知道子元素才能够推导出父元素。

反侵权盗版声明

　　电子工业出版社依法对本作品享有专有出版权。任何未经权利人书面许可，复制、销售或通过信息网络传播本作品的行为；歪曲、篡改、剽窃本作品的行为，均违反《中华人民共和国著作权法》，其行为人应承担相应的民事责任和行政责任，构成犯罪的，将被依法追究刑事责任。

　　为了维护市场秩序，保护权利人的合法权益，我社将依法查处和打击侵权盗版的单位和个人。欢迎社会各界人士积极举报侵权盗版行为，本社将奖励举报有功人员，并保证举报人的信息不被泄露。

举报电话：（010）88254396；（010）88258888
传　　真：（010）88254397
E-mail：　　dbqq@phei.com.cn
通信地址：北京市万寿路 173 信箱
　　　　　电子工业出版社总编办公室
邮　　编：100036